The High School For

Arts

&

Business

Ana R. Zambrano-Burakov
Principal
105-25 Horace Harding Expwy, N.
Corona, NY 11368
Phone: (718) 271-8383 x162
Fax: (718) 271-7196

Created by:

Leadership Acceleration Summer Program B.H.R.A.
(E.R.T.R.G.)
Bob Diamond
Greg Castillo

The Interns:
Moneshae Bailey
Katherine Trinidad

DATED: August 8, 2017

Table of Contents

Project Background

    NY Daily News July 19, 2017

    NY Daily News March 06, 2017

Preface

    NY Population Projection 2016-2045

    Ratio of Required Mass Transit Capacity Growth Rate to Population Growth Rate to (2:1)

    Lower East Side (LES) Mulberry Street Photo

    Three Part Transit Network Hierarchy

    The Influence of Mass transit on Economic Development and the Substantive Increase of Affordable Housing Units

    Residential and Commercial Development Patterns Follow Rail Transit Routes

    New York City population Density Map 2010 (Lets change blue to green)

Selection of Transit Mode

SOURCE:
Pages 100 and 101 of Vukan R. Vuchic's book "Urban Public Transportation Systems and Technology". Note that "RB" is Bus, "SCR" is "Streetcar", "SRB" is "Bus Rapid Transit", "RRT is Subway "Rail Rapid Transit", and of course, "LRT" is light rail

Conclusion: Rail Rapid Transit if the only mass transit mode that can handle "NYC Sized" major scale residential and commercial development

Route Alternatives

Comparison of Existing Junction Point Options

Diagrams of Existing Track Beds

Opportunity for Large Scale, Truly Affordable Housing development Along Rail Routes

Generalized Architectural Rendering of Skyscraper Residential Buildings

Potential Funding Sources

Cost Estimate Circa 2003

Cost Benefit Analysis
    Cost
    Benefits

Other Funding Examples

DAILY NEWS | OPINION

A rail good idea for Queens: Restore the Rockaway Beach line

EDITORIALS

NEW YORK DAILY NEWS   Wednesday, July 19, 2017, 5:49 PM

Put a train back here. (JEFF CHIEN-HSING LIAO FOR FRIENDS OF THE QUEENSWAY)

Halfway through his week governing from Queens, Mayor de Blasio owes the people of the borough and city his support for restoring rail service to the old Rockaway Beach Branch of the LIRR, a route that has been inactive since 1962.

Although it has been more than half a century since the last train ran the 3.5 miles from Ozone Park up to Rego Park, connecting to the LIRR Main Line and straight into Penn Station, the valuable and irreplaceable right of way remains.

Some activists want this fallow rail bed to become a park, an outer-borough High Line, called Queensway. Sorry, folks. We love the High Line as much as anyone and were the very first to champion it, before Michael Bloomberg had ever heard of the defunct elevated freight spur.

But the derelict High Line had only two possible futures: demolition or a park, modeled on the original greened viaduct, the Promenade Plantée in Paris. A return to service for either goods or passengers was never viable.

In contrast, the Rockaway Beach Branch is ideal for a fast, direct link from JFK and Southeast Queens into the heart of Manhattan. It was such a good idea that then-Gov. Nelson Rockefeller pushed it a half century ago.

De Blasio, when asked last month, said "We are closing in on our final decision. I want everything on the table. I will come back publicly with an assessment of the different options and the cost and we will move to a decision."

For a guy who is planning a trolley line along the Brooklyn/Queens waterfront and expanding ferry service, the call is an easy one: transit.

Just look at the subways. New York can't exist if New Yorkers can't move and a train on the Rockaway Beach Branch can move more people faster and more efficiently than any other mode. JFK could finally have a direct one-seat train ride into Midtown. And the most expensive part — acquiring the land — is already done, as the city owns the right of way.

Of course, reactivating the line for the subway or LIRR would require cooperation with Gov. Cuomo. De Blasio can deal with that.

The other choice, a linear park, should be just as easy to dismiss for a man like de Blasio who has never once even been to the High Line (too Bloombergian) . For now, the High Line will have to suffice for New Yorkers. Queens needs transit. Queens needs the Rockaway Beach Branch.

☐ Send a Letter to the Editor

The right train to the plane: JFK needs a one-seat ride

EDITORIALS
NEW YORK DAILY NEWS   Monday, March 6, 2017, 4:10 AM

Too little a train. (TURNBULL, BILL/BILL TURNBULL)

It was 20 years ago when the Dutch managers of Amsterdam's Schiphol airport — one of the world's most innovative and pleasant — landed in Queens to rehab JFK's Terminal Four.

On a visit to the terminal, we asked them if they were going to import to JFK Schipol's famous etchings of an insect in all men's room urinals. The fake fly was found to produce better aim and cleaner restrooms.
Of course, said the Dutchman, "don't people pee the same way here?"

And so the flies fly at Terminal Four.

The flying Dutchman also noted that while Schiphol and JFK handle the same passenger load, efficient Amsterdam had a single central terminal, while chaotic JFK had nine terminals spread out.

That sprawling mess is finally getting fixed, too — thanks to Gov. Cuomo's plans for a $10 billion tail-to-snout airport overhaul.

But Cuomo has still one more thing to copy from our Dutch friends: A direct, swift, one-seat rail ride to the city's center.

He says he wants to make it happen. Good luck with that, sir. We truly hope you succeed.

Since 2003, JFK has had something called the AirTrain. Running from Jamaica and Howard Beach, where it links up with subway stations, it's nowhere to a one-seat ride. That makes trips to the airport far too long, if you take public transit — or

far too pricey, if you take a cab or car service.

There's an easy answer. When he was governor in 1968, Nelson Rockefeller wanted to reactivate the closed Rockaway Beach Long Island Rail Road line for a one-seat into Penn Station — then as now the best route. It didn't happen.

Two decades later, Mario Cuomo and Sen. Pat Moynihan tried again. It didn't happen.

Another decade went by. Frustrated by buck-passing between the Port Authority and MTA, we invited brass from both in for a meeting.

They explained that a one-seat ride was a no-go because the AirTrain, LIRR and subway cars couldn't, for various technical reasons, run on each other's tracks. But a hybrid, called the "fourth car," could operate on all three. It just had to be designed, at an estimated cost back then of $50 million. Not too bad.

Create that — eminently possible given late-90s technology, and even easier to pull off now — and run it over existing rail lines between Manhattan and Queens. Voilà, a true one-seat ride.

The MTA and PA men acknowledged this was technically doable and could provide ideal service at almost all times, though during the peak of the weekday rush it wouldn't be quite as frequent.

And, hung up on that asterisk, we've had no one-seat ride since the AirTrain opened in 2003.

To get the one-seat ride he wants and New York deserves, Cuomo must force the MTA and PA to work up blueprints for the fourth car immediately. He also needs to restore the Rockaway Beach line — the ideal route for the train to the plane — for transit, killing the silly dream of making it a Queens version of Manhattan's High Line.

We love the High Line and were among its first backers, but Queens desperately needs better transit.

Cuomo, returning from Israel today, lands at JFK. He must follow the smart Schiphol model and be the one to finally build the one-seat ride.

Stadler Rail AG

Type

Limited company

Industry
Locomotive engineering

Founded
1942/1997 (Holding)

Headquarters
Bussnang, Switzerland

Key people
Peter Spuhler
(CEO and Vice President)

Revenue
2.2 billion Swiss francs (2012)

Number of employees
6,000 (2012)[1]

Website
www.stadlerrail.com

GTW

The GTW, an abbreviation of the German for articulated multiple-unit train, is a single-decker regional train. With the first generation of its own articulated multiple unit vehicles, Stadler laid the foundations in 1995 for its current success as a train builder with a comprehensive passenger transport package.

The articulated multiple unit is a low-floor single-decker regional train. The tractive unit is located between the carriages, although the train is still accessible throughout. Depending on their design, the articulated multiple unit trains achieve a maximum speed of 140 km/h. **They are available in train sets of two to four cars, in narrow, standard and broad gauge, as an electric multiple unit EMU), diesel multiple unit (DMU) or bimodal multiple unit (BMU).**

SUSTAINABILITY

FRA Approves First Integrated Use of Stadler GTW Rail Vehicle for DCTA

SOURCE: DENTON COUNTY TRANSPORTATION AUTHORITY (DCTA) JUN 5, 2012

Denton County Transportation Authority's A-Train at the Trinity Mills Station.

Photo credit: *Photo courtesy of DCTA.*

On Monday, June 4, 2012, Administrator Joseph Szabo of the Federal Railroad Administration (FRA) in conjunction with the American Public Transportation Association Annual Rail Conference formally announced approval of DCTA's request to operate the Stadler GTW concurrent with traditional, compliant equipment. This means that for the first time ever; light-weight/fuel efficient, eco-friendly low-floor vehicles will be permitted to operate in rail corridors concurrently with traditionally compliant vehicles. The waiver, a first of its kind, will expand commuter rail options for transportation authorities across the United States.

In 2009, the FRA's Rail Safety Advisory Committee (RSAC) prepared a set of technical criteria and procedures for evaluating passenger rail train-sets that have been built to alternative designs. The alternative designs enable lighter, more fuel-efficient rail vehicles equipped with a Crash Energy Management system to commingle with traditionally compliant equipment. The DCTA/Stadler alternative design waiver is the first comprehensive submittal that follows the RSAC Engineering Task Force (ETF) procedures for Tier I equipment. The approval of the DCTA/Stadler waiver request demonstrates that the enhanced crashworthiness and passenger protection systems inherent to DCTA's new rail vehicles meet the latest and most stringent safety standards in the U.S.

"Stadler is excited and proud to have the opportunity of announcing this milestone and appreciates the immense joint effort conducted by DCTA and the FRA," stated Steve Bonina, Stadler USA president. "Stadler continues to be hopeful that the FRA codifies the RSAC guidelines into regulatory requirements in order to open the North American Rail Network to this outstanding, safer, eco-friendly rail technology, which will help to make rail systems safer, more efficient, more reliable and less costly."

Stadler, DCTA and DCTA's vehicle consultant, LTK Engineering Services, have been working closely with the FRA to achieve this waiver since 2009. DCTA partnered with Stadler to make modifications and enhancements to the GTW to comply with the required safety guidelines. Modifications include changes to the fuel tank design, window glazing and passenger and operator seats.

"This approval is the result of unprecedented cooperation between DCTA, the FRA LTK and Stadler," stated Jim Cline, DCTA president. "Our efforts to operate the nation's first alternative compliant vehicle demonstrate not only our commitment to increased safety for our passengers and operators but to improving safety for the commuter rail industry. We are setting the conditions for future success for commuter rail expansion in North Texas and the conditions will allow us to advance the integration of these vehicles onto our system."

DCTA purchased 11 diesel-electric GTW 2/6 articulated rail vehicles from Stadler. The vehicles are compliant with the Americans with Disabilities Act (ADA), and incorporate enhanced air conditioning, passenger information system, video surveillance and numerous FRA compliant elements. The spacious interior has room for wheelchairs, strollers and bicycles. There are 104 seats and standing room for 96 persons in every vehicle along with bright compartments, large windows and comfortable seating. DCTA and Stadler will begin integrating the cars into service this summer.

Voice your opinion!

No comments have been added yet. Want to start the conversation?

This site requires you to login or register to post a comment.

NEWS

CA: California Supreme Court Upholds Cap-and-Trade System

June 29, 2017

California's cap-and-trade law, which requires companies to buy permits to emit climate-changing greenhouse gases into the air, survived a legal challenge Wednesday when the state Supreme Court turned down an appeal by business groups.

PRESS RELEASE

L.A. Metro Plans Activities in Support of Bike Month this May in Los Angeles

May 18, 2017

From **LOS ANGELES COUNTY METROPOLITAN TRANSPORTATION AUTHORITY (METRO)**

The Los Angeles County Metropolitan Transportation Authority (Metro) has scheduled a number of activities this May to commemorate Bike Month

PRESS RELEASE

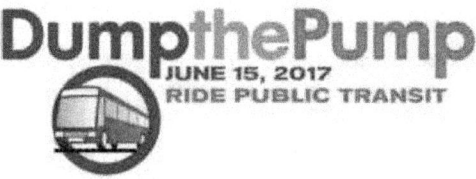

Laketran Celebrates Dump the Pump Day, June 15

May 23, 2017
From **LAKETRAN**

With the fluctuation of summer gas prices and need to weaken dependence of foreign oil, Laketran, along with public transportation systems nationwide, is participating in the 12th Annual National Dump the Pump Day.

PRESS RELEASE

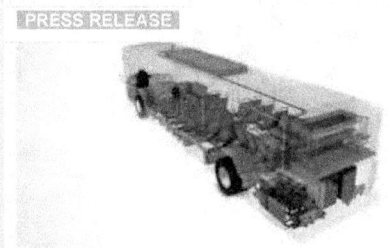

Preface

- The latest U.S. Census statistics show the City of New York has grown by 375,000 people in **the past 5 years***. The affordable housing crisis and the city's overburdened transportation infrastructure are only going to be further exacerbated.

At the current rate, we're looking at a population of *10.1 million New Yorkers by 2036*:

- 8.6 million in mid 2016 (current)
- 9 million by 2021
- 9.35 million by 2026
- 9.725 million by 2031
- 10.1 million by 2036

> **Note: Some estimates place the current NYC population at roughly 9 million, and project Brooklyn's population alone will nearly double to well over 5 million by 2040, with a total NYC population of over 12 million - and possibly rising as high as 15 million people.**

Sources:

http://newyorkyimby.com/2015/04/new-york-city-is-already-at-its-2020-population-forecast.html

http://www.ny1.com/nyc/all-boroughs/news/2016/03/23/nyc-s-population-tops-8-5-million-for-first-time--new-census-figures-show.html

http://gothamist.com/2016/08/09/oh_hello_brooklyn.php

Article II.

With an increase in population from 1910 to 1920 of 35%, the increase in New York subway and elevated traffic may be estimated tentatively at 70%. These long distance estimates must be accepted as provisional, but the results arrived at are based upon the best figures now obtainable. The ratio of 2 to 1 for the increase in traffic compared with the population gain is generally accepted as conservative by traffic engineers who have made a specialty of the New York situation. It was also just about the ratio actually shown in the decade 1900-1910 as will be seen from the following table:

	Increases		% Increases	
	Population	Traffic	Population	Traffic
1860-1870	303,324	97,753,000	25.8	192.3
1870-1880	433,595	138,918,000	29.3	94.5
1880-1890	695,716	312,913,000	36.4	108.8
1890-1900	829,788	245,939,000	31.8	40.9
1900-1910	1,329,681	684,909,000	38.7	80.9

TRAFFIC GROWTH TWO AND THREE TIMES POPULATION GROWTH.

Covering 50 years from 1860 to 1910 the average increase in population, by decades, has been 32.4% while the average gain in traffic has been 103.5%. Thus, traffic gains have averaged three times population gains.

Figure 1. Three-part transit network hierarchy (20).

LEVEL ONE

Good rush hour service to Central Business District

LEVEL TWO

Good service to Central Business District at all times

LEVEL THREE

Good service throughout metropolitan region

Source:
Light Rail Transit
TRB Special Report 161, 1975

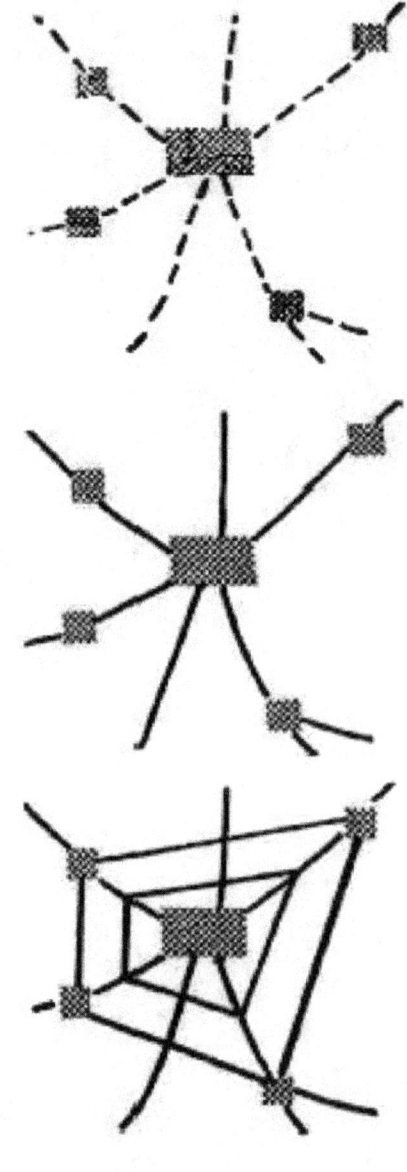

New York's Changing Scene

NEWS PHOTO BY DANIEL JACINO

In 1917 (top photo), Queens Boulevard was a narrow thoroughfare as it cut across Long Island City past the Rawson St. station on the IRT Main St. Flushing line. Acres of barren land awaited the industrial boom that was to follow World War 1. Now (bottom), the area, just minutes from Manhattan, is a concentration of small manufacturing establishments.

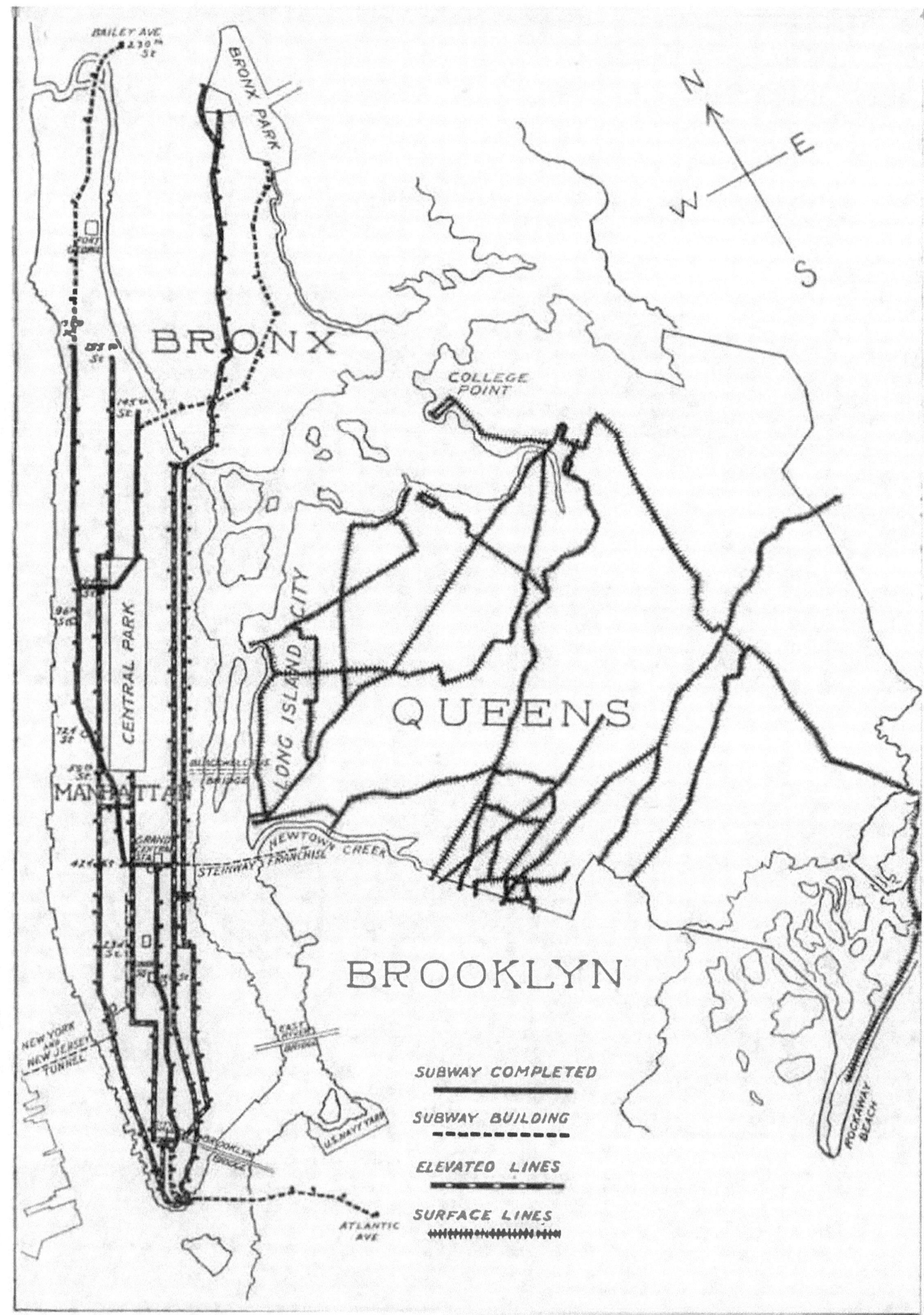

New York City in 3D

Census tract population density, 2010

Population - 8,175,133
Source: 2010 Census

- > 100,000 people per square mile
- > 80,000 people per square mile
- > 60,000 people per square mile
- > 40,000 people per square mile
- > 20,000 people per square mile
- 0 - 20,000 people per square mile

in 2D

New York City is the most densely populated city in the United States, with an average density of just under 28,000 people per square mile. In much of Manhattan and the Bronx, and parts of Queens and Brooklyn, the population density for some census tracts is higher than 100,000. Staten Island only has one census tract with a density above 60,000 but in comparison to most of the United States it is still very densely populated.

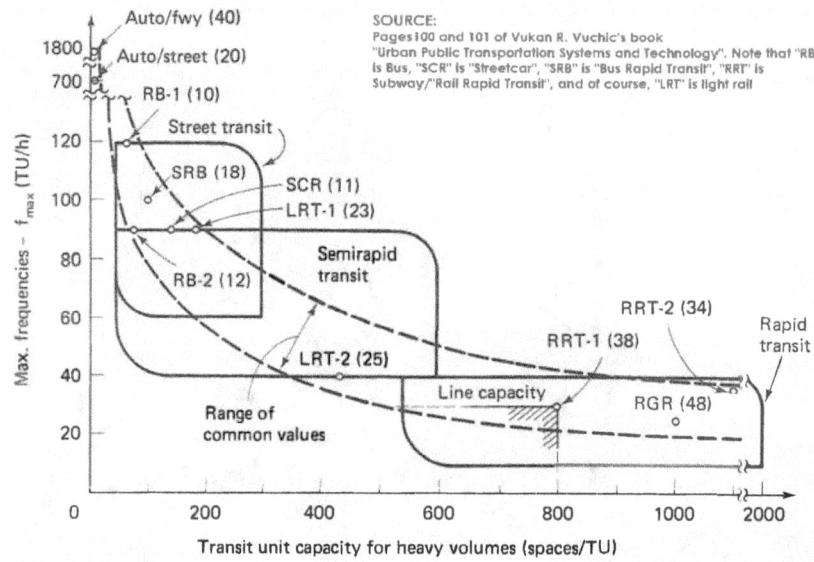

Figure 2.18. Vehicle capacities, maximum frequencies and line capacities of different modes

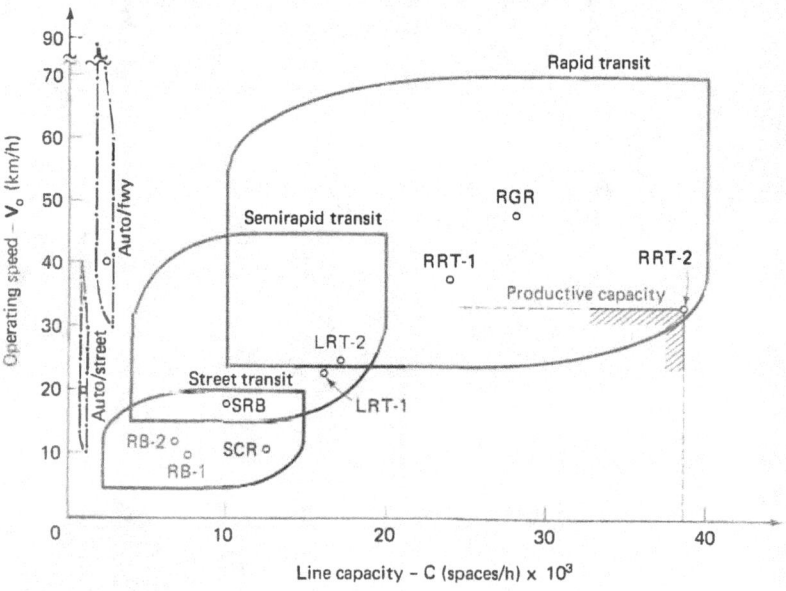

Figure 2.19. Line capacities, operating speeds, and productive capacities of different modes

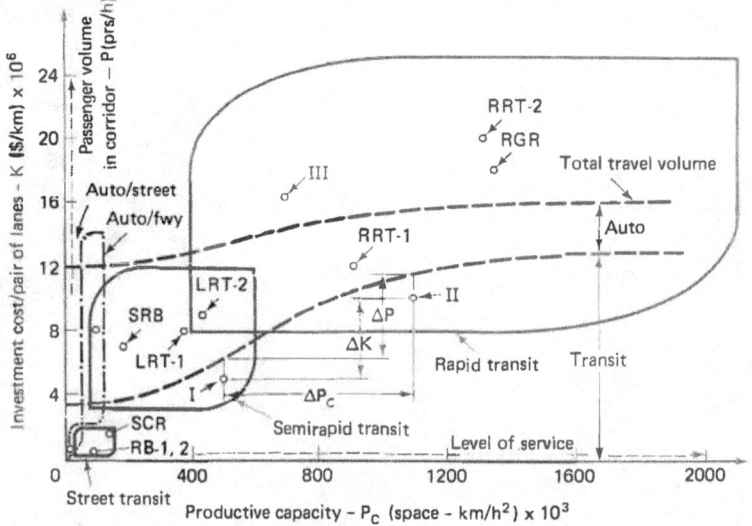

Figure 2.20. Relationship between productive capacity, investment cost, and passenger attraction of different generic classes of transit modes

From Vucan Vuchic's book

TRANSPORTATION WORKSHOP

Technical, Operational, and System Characteristics of Urban Public Transportation Modes

CHARACTERISTICS	MODE/UNIT	STREET TRANSIT		SEMIRAPID TRANSIT		RAPID TRANSIT
		RB	SCR	SRB	LRT	RRT
1. Vehicle capacity, C_v	sps/veh	40-120	100-180	40-120	110-250	140-250
2. Vehicle/transit unit	veh/TU	1	1-3	1	1-4	1-10
3. Transit unit capacity	sps/TU	40-120	100-300	40-120	110-600	140-2000
4. Maximum technical speed, V	km/h	40-80	60-70	70-90	60-100	80-100
5. Maximum frequency, f_{max}	Tu/h	60-120	60-120	60-90	40-90	20-40
6. Line capacity, C	sps/h	2,400-8,000	4,000-15,000	4,000-8,000	6,000-20,000	10,000-40,000
7. Normal operating speed, V_o	km/h	15-25	12-20	20-40	20-45	25-60
8. Operating speed at capacity, V_c	km/h	6-15	5-13	15-30	15-40	24-55
9. Productive capacity, P_c	(sp-km/h^2) x10^3	20-90	30-150	75-200	120-600	400-1800
10. Lane width (one-way)	m	3.00-3.65	3.00-3.50	3.65-3.75	3.40-3.75	3.70-4.30
11. Vehicle control	-	Man./vis.	Man./vis.	Man./vis.	Man./vis.-sig.	Man.-aut./sig.
12. Reliability	-	Low-med.	Low-med.	High	High	Very high
13. Safety	-	Med.	Med.	High	High	Very high
14. Station spacing	m	200-500	250-500	350-800	350-800	500-2000
15. Investment cost per pair of lanes	($/km)x 10^6	0.1-0.4	1.0-2.0	3.0-9.0	3.5-12.0	8.0-25.0
	$/ft	30-122	305-610	914-2743	1067-3658	2438-7620

RB ↑ Street Bus
SCR ↑ Street Trolley
SRB ↑ Busway
LRT ↑ Trolley in Exclusive Lane
RRT ↑ Heavy Rail

Graph shows light rail fills capacity gap between bus and subway.

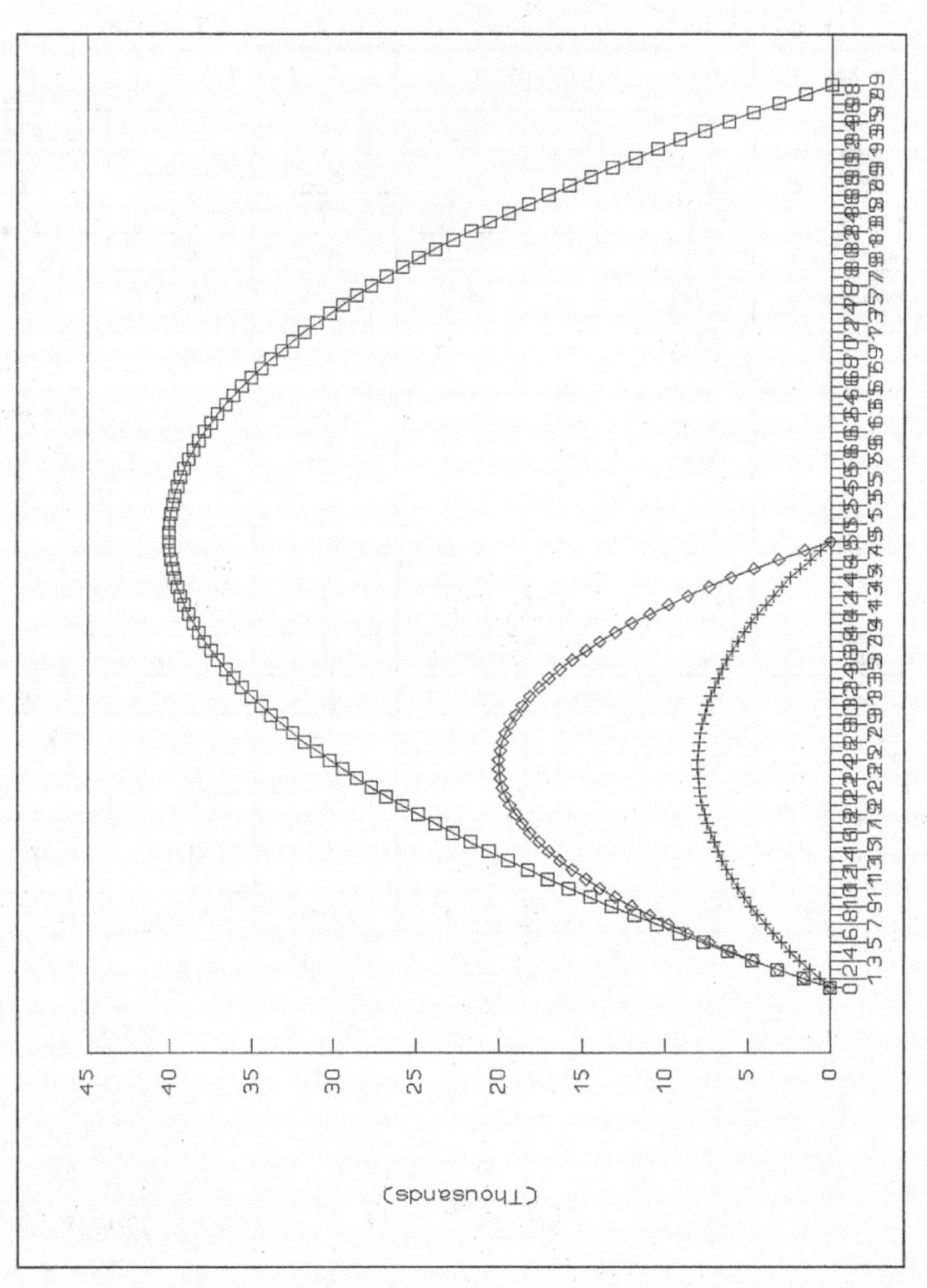

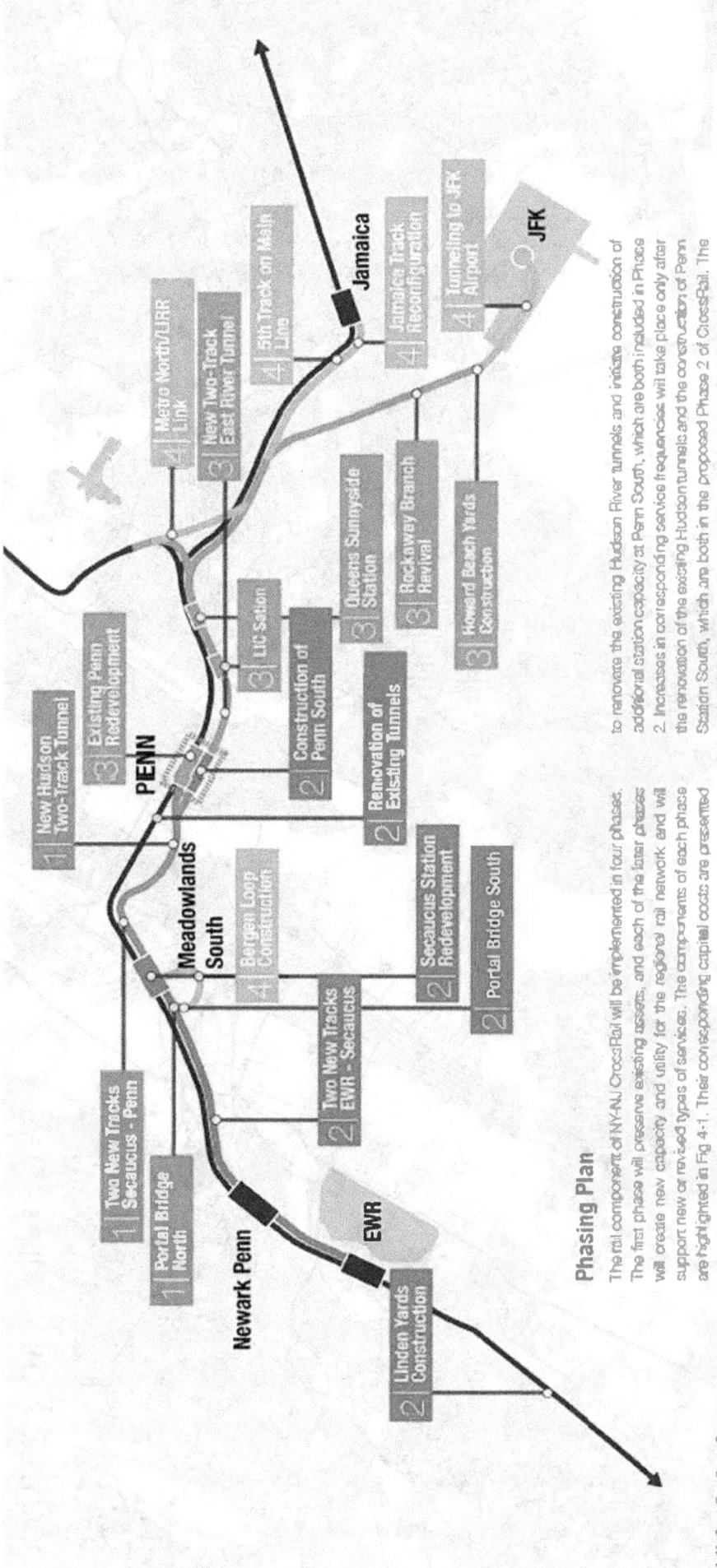

[Fig. 4-1] CrossRail Phase Plan

Phasing Plan

The rail component of NY-NJ CrossRail will be implemented in four phases. The first phase will preserve existing assets, and each of the later phases will create new capacity and utility for the regional rail network and will support new or revised types of services. The components of each phase are highlighted in Fig 4-1. Their corresponding capital costs are presented in the Finance Section, in Fig 6-1.

to renovate the existing Hudson River tunnels and initiate construction of additional station capacity at Penn South, which are both included in Phase 2. Increases in corresponding service frequencies will take place only after the renovation of the existing Hudson tunnels and the construction of Penn Station South, which are both in the proposed Phase 2 of CrossRail. The completion of East Side Access with direct LIRR service to Grand Central, will free up slots at Penn Station by rerouting some LIRR trains to the

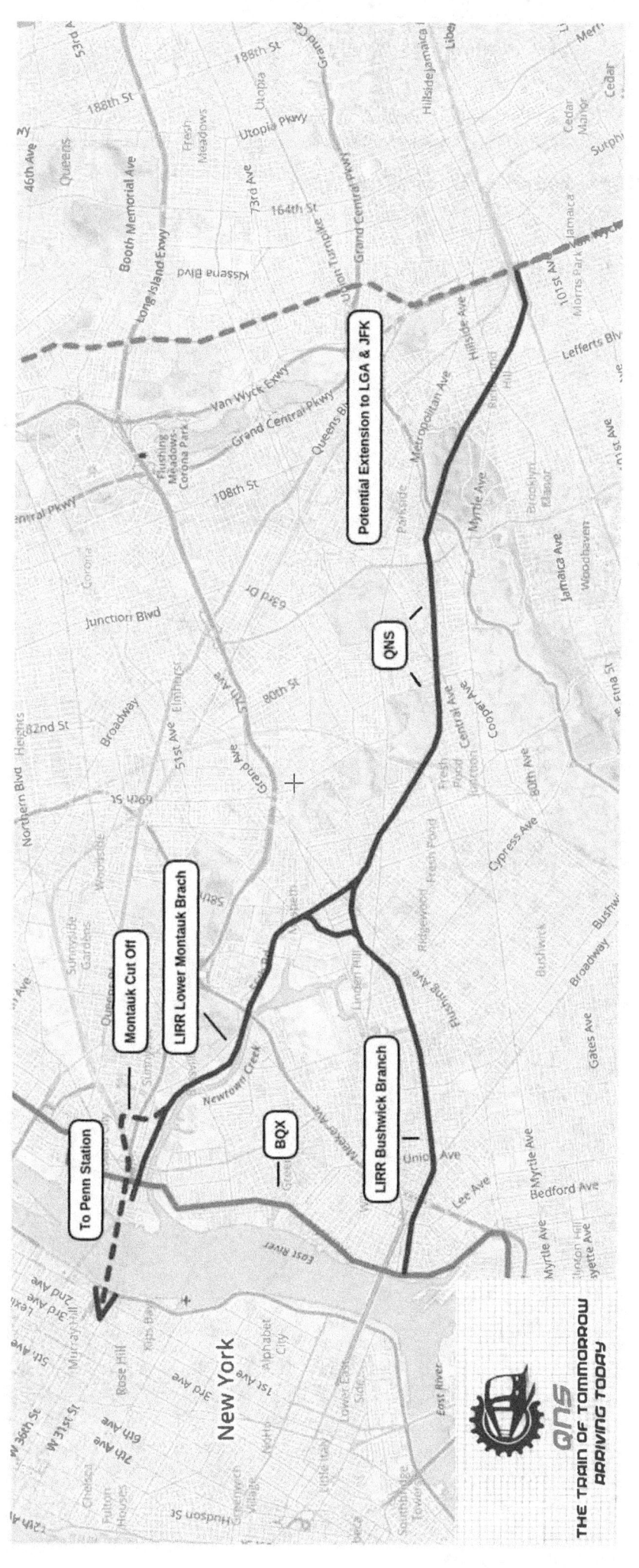

Subject: JFK- Midtwn Exp- VIA RBB, Lower Montauk, Montauk Cut Off, Sunnyside Loop, to Penn Tubes

From: Robert Diamond (rdiamond@brooklynrail.net)

To:

Date: Saturday, April 1, 2017 8:27 PM

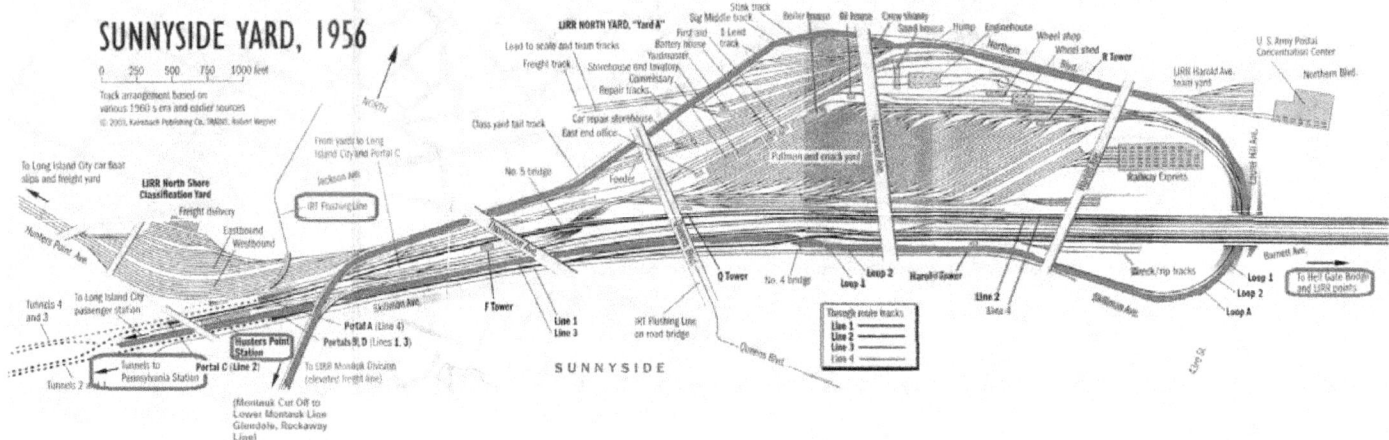

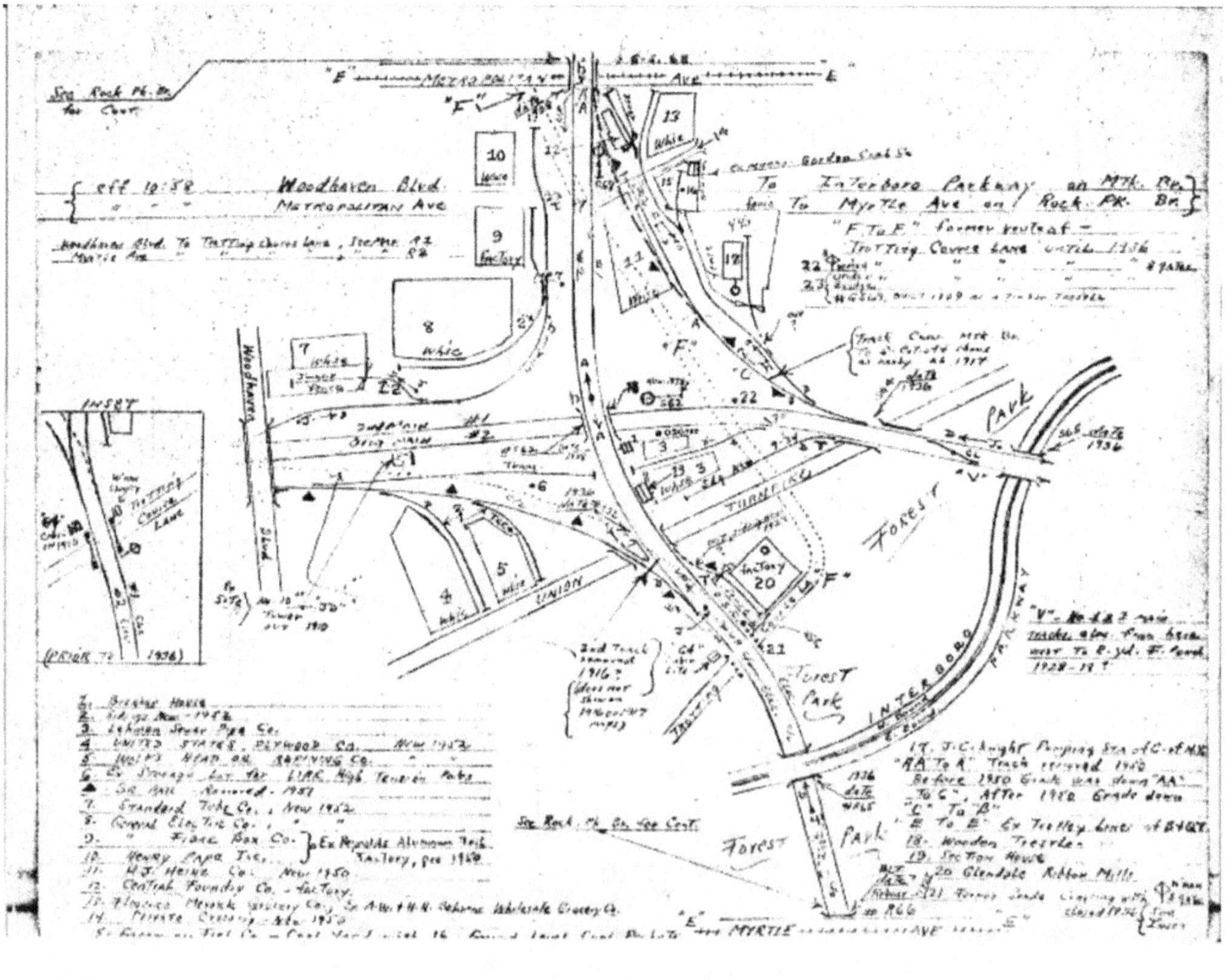

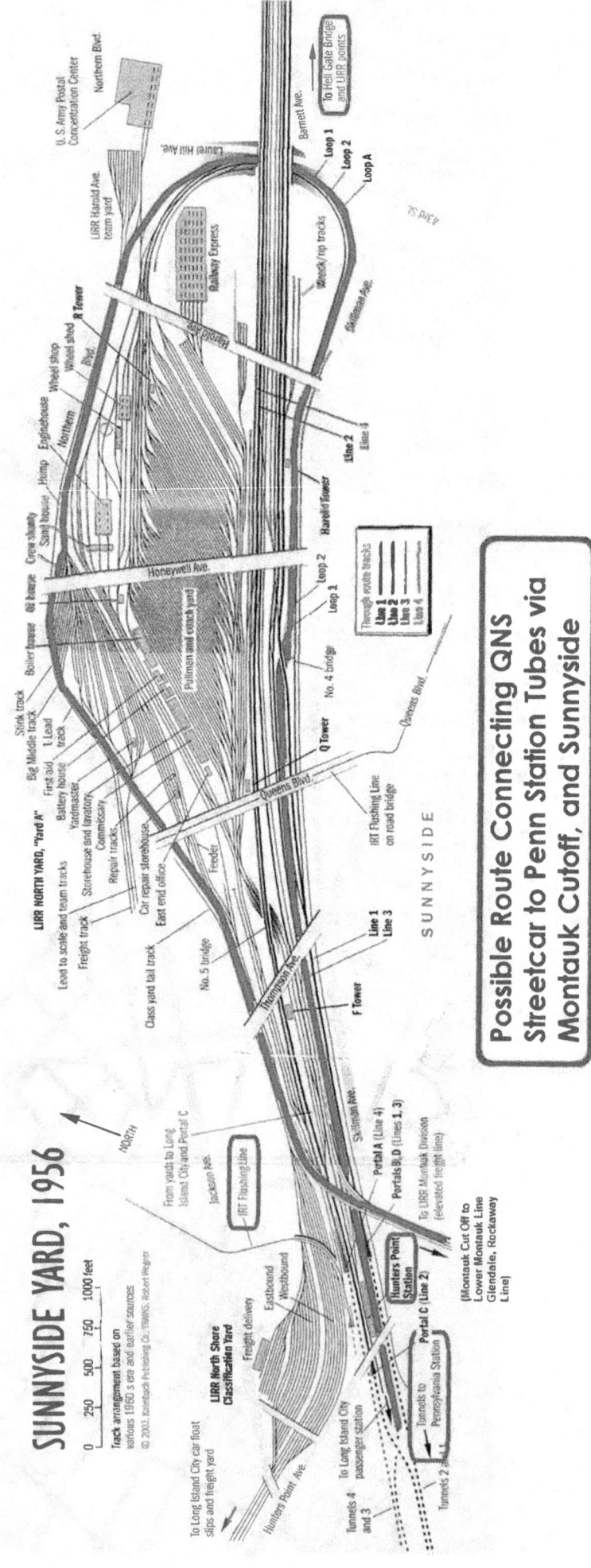

Montauk Cutoff

On the extreme northwest (railroad south) end of the platforms, high turnstiles lead to a single staircase that goes up to either western corners of 63rd Road and Queens Boulevard, the northwest one for the Manhattan-bound platform and the southwest one for the Forest Hills-bound platform.

Unfinished Rockaway spur

East of this station, there is an unfinished signal tower on the Jamaica-bound (railroad north) platform and a bellmouth that diverges to the south from the local track. Another bellmouth from the Manhattan-bound local track diverges north, then curves south above the Queens Boulevard Line to join the other bellmouth. These were provisions for a planned expansion in the 1930s that would have connected with the IND Rockaway Line (formerly a Long Island Rail Road branch) towards Howard Beach, JFK Airport, and the Rockaways.[5][6][7][8] This spur would have run down 66th Avenue before joining the Rockaway Line at its former junction with the LIRR Main Line.[7] In January 2013, a petition was started on change.org to make use of the bellmouths to connect the station to the currently unused portion of the Rockaway Line.[9]

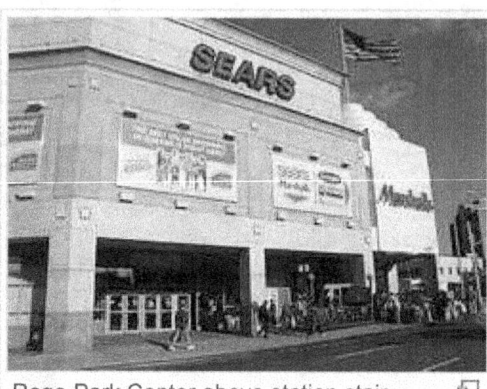
Rego Park Center above station stair

References

Cite error: Invalid `<references>` tag; parameter "group" is allowed only.

Use `<references />`, or `<references group="..." />`

External links

- nycsubway.org— IND Queens Boulevard Line: 63rd Drive/Rego Park
- nycsubway.org— The History of the Independent Subway:
- Station Reporter — R Train
- Station Reporter — M Train
- Forgotten NY: Subways and Trains — Rockaway Branch
- Forgotten NY: Subways and Trains — Subway Signs to Nowhere
- The Subway Nut - 63rd Drive – Rego Park Pictures
- 63rd Road exit only stair from Google Maps Street View
- 63rd Drive entrance from Google Maps Street View
- 64th Road entrance from Google Maps Street View

Wikimedia Commons has media related to *63rd Drive – Rego Park (IND Queens Boulevard Line)*.

v·t·e	New York City Subway stations	[show]

Proposed connection (1940) linking the LIRR Rockaway Line to the IND subway at 66th ~~Street~~ (Avenue) and Queens Boulevard in Rego Park.

Sorry if this attached file is too big -- hope you can view it.

Comments are welcome!!! I've heard rumors the flying junction for this connection actually exists under Queens Boulevard, but aside from the work done in the actual width of the subway route, there is no additional tunneling between the Rockaway Line and the subway line.

Talk about a route for the train to the plane!

(Yes, I copied this map off of an "eBay" auction page -- the price is over $50 so far.)

Bernie

THINKING BIG

Source: Crain's NY Business [Circa 2003]

Major transit projects proposed for the New York region:

PROJECT	ESTIMATED COST	LEAD SPONSOR/BACKER
Rebuilding lower Manhattan		
New PATH station and new downtown concourse	$2 billion	Port Authority
New Fulton Street transit center	$750 million	MTA
Reconfiguration of South Ferry subway station	$400 million	MTA
WTC bus station and West Street platform	$500 million to $1.65 billion	Bloomberg/Port Authority
East Side Access	$5.3 billion	MTA
Second Avenue Subway		
Northern segment (125th Street to 63rd Street)	$6.5 billion	MTA
Lower segment (63rd Street to downtown)	$6 billion	MTA
MetroLink (downtown to Brooklyn)	$3 billion	RPA
No. 7 subway line extension	$1.5 billion	Bloomberg
Farley Post Office	$1 billion	Former Sen., Daniel Patrick Moynihan
JFK: one-seat ride to midtown	$700 million	NY econ. development czar Charles Gargano
JFK: one-seat ride to downtown	$3.7 billion	Bloomberg
Super subway shuttle	$2 billion to $5 billion	Brookfield Financial Properties
Newark airport: one-seat ride to downtown	$775 million	Port Authority
Access to the Region's Core	$4.5 billion	NJ Transit
Cross-harbor rail freight tunnel	$4 billion to $7 billion	Nadler
Tappan Zee Bridge rebuilding	$4 billion	NYS Thruway Authority
Gowanus Expressway	$1 billion to $2.5 billion	NYS DOT
Total estimated cost	**$47.625 billion to $56.275 billion**	

Sources: agencies, news reports

SEARCHING FOR DOLLARS

Potential sources of revenue:

SOURCE	AMOUNT	LEAD SPONSOR/BACKER
Tolls on East River bridges	$8 billion	
Renewal of Transportation Equity Act for the 21st Century	$5 billion[1]	Port Authority
FEMA/U.S. DOT funds for downtown	$4.55 billion	MTA
WTC lessee insurance	$2.94 billion	MTA
Tax increment financing (far West Side)	$1.5 billion	Bloomberg/Port Authority
Lower Manhattan Development Corp. funds	$1.25 billion	MTA
Excess Port Authority funds	$940 million	
Passenger facility charge	$830 million	MTA
NYC capital budget	$106 million[2]	MTA

1-Estimated funding over six years. 2-Annual contribution to MTA's capital budget.

Sources: agencies, news reports

Sky Scraper Residential Development Strategically Located Along New Rail Routes

From:

To:

Sent: Thu, Jul 20, 2017 12:59 PM

Subject: Affordability in NYC

NOTE: THIS CAMPAIGN AD IS CITED FOR TECHNICAL INFORMATION PURPOSES ONLY

As I talk to people throughout our city, the first issue I hear from almost everyone I meet is how difficult it can be to afford to live here.

That's why we launched the most ambitious affordable housing plan in the history of New York City—more than $40 billion to **build and protect 200,000 affordable apartments by 2024.** That effort will help more than 500,000 New Yorkers afford to stay here.

I wrote an op-ed in the *New York Daily News* last week about the work we're doing to tackle the affordability crisis in our city. I hope that you'll read it and forward it to your friends.

Building a More Affordable City

By: Bill de Blasio

In a lot of ways, New York isn't the city I moved to back in 1979.
I'm old enough to separate my nostalgia for those days from the reality of how dangerous and uncertain they could be. Today, our streets are safer than ever. Schools are improving. Unemployment is at record lows.

But in one fundamental way, we risk becoming victims of our own success. As more parents stay here to raise their kids, more seniors retire in their neighborhoods instead of the Sun Belt, and more young people and immigrants come here because this is where the good jobs can be found, the pressure on our city's housing stock has led to a full-blown affordability crisis.

When we took office in 2014, half of renters were spending more on housing than they could afford. Rents had risen sharply through the recession, even as incomes were tumbling. The rising anxiety built up over decades hit a fever as hundreds of thousands of New Yorkers asked themselves: Can I still afford to live here?

We heard and felt the pleas, and responded with the biggest affordable housing boom in the country's history — a $41 billion plan to build and protect 200,000 affordable apartments by 2024, enough to help half a million New Yorkers afford to stay here.

And to the surprise of the skeptics and naysayers, we've actually delivered on our promise. On Thursday, we're announcing that the city government has financed nearly 78,000 newly constructed or protected affordable homes since 2014. That's the most affordable housing produced in any three years in New York City's history — enough housing for the entire population of Salt Lake City, financed in just three years' time.

We are ahead of schedule and on budget, and the pace is accelerating, with the fiscal year that just ended on June 30 our biggest yet: more than 24,000 homes.

These apartments aren't just expanding the housing supply; they directly address the affordability problem with tightly regulated rents governed by the city, tied to the incomes of their tenants. Households typically pay no more than 30% of their income in rent.

What we've gotten right here in New York is a lesson for the whole country as we face an attack on federal affordable housing programs from the Trump administration and congressional Republicans.

Heavy Rail: Line & System Average Scheduled Speed

System Average	System	Line	Speed
17.4	New York City Subway	Franklin Park S	13.3
		C	13.5
		M	13.7
		W	14.3
		L	15.3
		R	15.3
		J	15.4
		Q	15.4
		1	15.5
		2	15.9
		6	15.9
		3	16.2
		D	16.7
		N	16.8
		C	16.9
		6	17.0
		V	17.0
		Z	17.4
		7	17.5
		B	17.9
		F	18.0
		4	18.1
		5	18.3
		A	19.0
		E	19.6
		SIRT	20.0
		◆	20.8
		Rockaway S	22.1
		42nd Street S	32.0
18.2	New York/New Jersey PATH	33rd - Journal Sq	14.8
		33rd - Hoboken	15.9
		Hoboken - WTC	18.0
		Newark - WTC	24.2
18.6	Philadelphia SEPTA	Broad Street	16.5
		Broad - Ridge	18.0
		Market-Frankford	19.8
		Broad Str Exp	21.5
19.4	Boston T	Red - Ashmont	17.8
		Blue	18.9
		Orange	19.4
		Red - Braintree	21.5
22.9	Chicago L	Blue	16.7
		Purple	17.4
		Brown	18.0
		Green - Cottage Gr	21.4
		Red	21.6
		Pink	23.2
		Green - 63rd	23.5
		Orange	25.9
		Yellow	38.3
24.9	Cleveland	Rapid	24.9
26.4	Los Angeles Metro	Purple	22.6
		Red	30.3
28.6	Miami Metro	Metrorail	28.6
28.8	Atlanta MARTA	Green	25.0
		Blue	28.1
		Red	30.9
		Gold	31.2
29.5	Washington Metro	Yellow	25.9
		Blue	27.9
		Green	29.3
		Red	31.4
		Orange	32.9
31.5	Baltimore Metro	Metro Subway	31.5
33.1	San Francisco BART	Fremont - Richmond	24.4
		Richmond - Millbrae	32.3
		Dublin - Daly City	35.2
		Pittsburg - SFO	36.4
		Fremont - Daly City	37.0
34.1	PATCO	Speedline	34.1

* All lines were measured based on their route structure and scheduled travel times at approximately 5:30P on a weekday. Analysis was conducted in February 2009.

Blackstone Unveils $40 Billion Infrastructure Mega Fund with Saudi Arabia as President Trump Visits

[5/21/2017 9:48:48 PM]

https://www.forbes.com/...17/05/20/blackstone-unveils-40-billion-infrastructure-mega-fund-with-saudi-arabia-as-trump-visits/#216215f67ad7

As President Donald Trump visits Saudi Arabia on the biggest foreign trip of his presidency, private equity giant Blackstone Group is unveiling a $40 billion infrastructure fund with the gulf nation, which will primarily invest in the United States. Saudi Arabia will commit $20 billion to the Blackstone infrastructure fund and another $20 billion will be raised from other limited partners, readying cash that could lead to $100 billion in total infrastructure investments on a leveraged basis. Blackstone, co-founded by billionaire Stephen Schwarzman, head of Trump's business council, said on Saturday morning the fund effectively launches a new business line for the over $360 billion in assets firm.

Already a powerhouse in private equity, real estate, hedge funds and credit investments, Blackstone has been preparing for a push into infrastructure, spotting an opportunity as large investors increasingly plant their money into the cogs of the global economy such as buildings, ports, wireless infrastructure, pipelines, railroads and airports. Through its various funds, Blackstone has invested over $40 billion in such investments over the past fifteen years; however, those deals were not made inside a dedicated infrastructure business. With large institutional investors and sovereign wealth funds spotting trillions of dollars in necessary infrastructure investments in the U.S. and globally in coming decades, they are looking to the sector as a green field area to earn stable double digit returns.

Firms like Global Infrastructure Partners, Macquarie, Brookfield Asset Management and Black Rock have already built weighty infrastructure businesses; now after lots of speculation Blackstone is entering the market in a major way.

Saudi Arabia's Public Investment Fund has agreed to commit $20 billion to the Blackstone infrastructure fund, which will be set up a permanent capital vehicle. Other outside investors will provide the rest of the fund's commitments. Though Blackstone says the vehicle "launches a new business" for the firms and is the culmination a year's worth of negotiation, it is being done on a non-binding basis, meaning talks continue and the structure has not been formalized. "There is broad agreement that the United States urgently needs to invest in its rapidly aging infrastructure. This will create well-paying American jobs and will lay the foundation for stronger long term economic growth," said Blackstone's billionaire president Hamilton E. James, in a press release announcing the deal.

"Blackstone has the talent, scale and experience to be an effective private sector partner in filling the massive infrastructure funding gap," James said, before characterizing Saturday's deal as a "vote of confidence in our country and Blackstone." H.E. Yasir Al Rumayyan, managing director of PIF, said Saudi Arabia's $20 billion commitment is a vote of confidence in President Trump's economic agenda. "This potential investment reflects our positive views around the ambitious infrastructure initiatives being undertaken in the United States as announced by President Trump, and the strategic opportunity for the Public Investment Fund to achieve long-term returns given historical investment shortfalls," Rumayyan said. "We look forward to partnering with Blackstone, a recognized leader with a strong record of achievement across its extensive infrastructure projects," he added. On Friday, PIF formally closed its commitment to SoftBank's $93 billion technology fund, underscoring its appetite for major new investments as Saudi Arabia diversifies its economy. Were one to count the $40 billion infrastructure fund towards Blackstone's assets under management, it would push overall AuM to over $400 billion, up roughly fourfold from its 2007 initial public offering. FORBES documented Blackstone's dramatic post-crisis rise in a May 2016 cover story on Schwarzman and the firm.

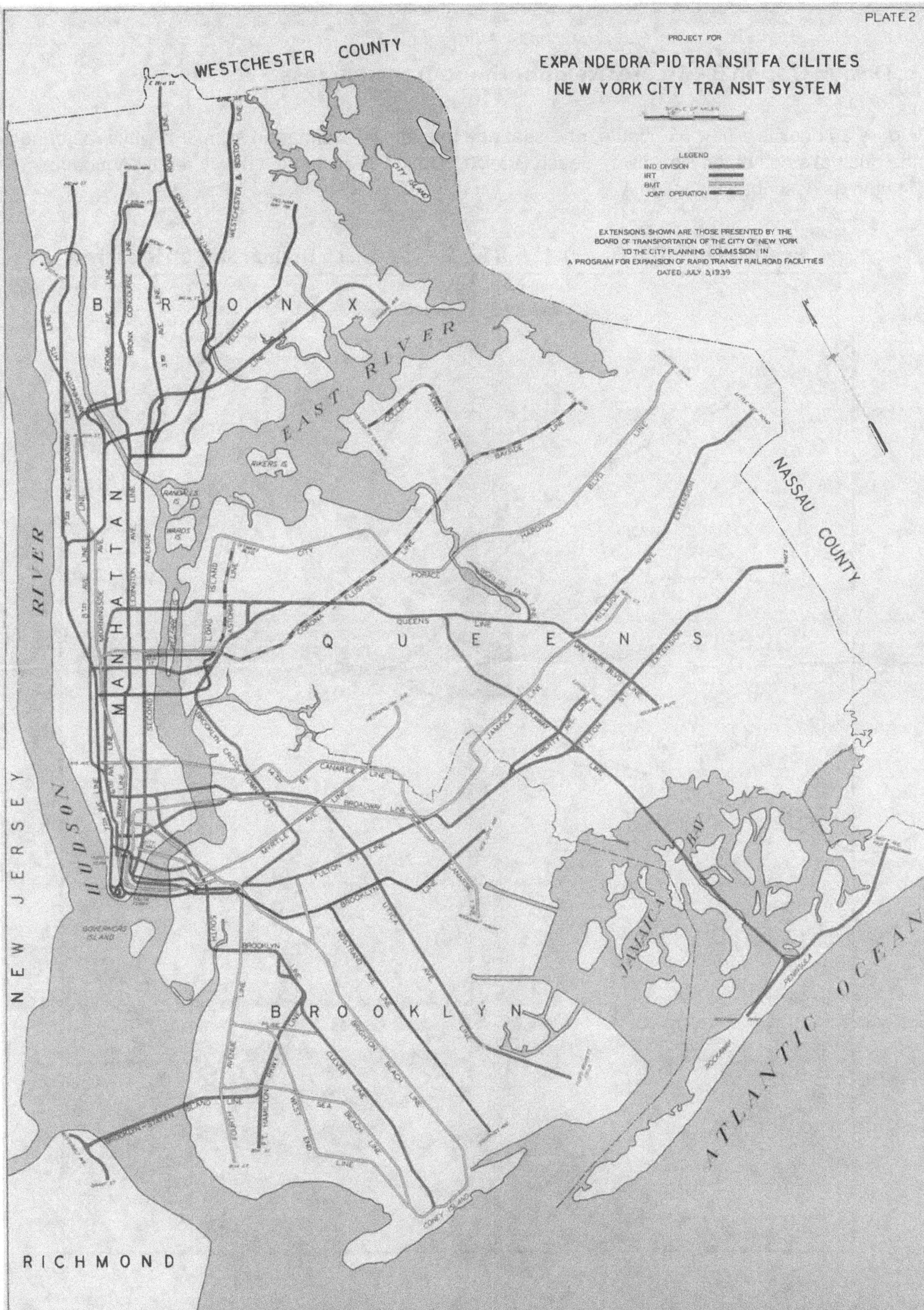
PLATE 2
PROJECT FOR
EXPANDED RAPID TRANSIT FACILITIES
NEW YORK CITY TRANSIT SYSTEM
LEGEND
IND DIVISION
IRT
BMT
JOINT OPERATION
EXTENSIONS SHOWN ARE THOSE PRESENTED BY THE
BOARD OF TRANSPORTATION OF THE CITY OF NEW YORK
TO THE CITY PLANNING COMMISSION IN
A PROGRAM FOR EXPANSION OF RAPID TRANSIT RAILROAD FACILITIES
DATED JULY 5, 1939
WESTCHESTER COUNTY
BRONX
EAST RIVER
MANHATTAN
QUEENS
NASSAU COUNTY
BROOKLYN
HUDSON RIVER
NEW JERSEY
RICHMOND
JAMAICA BAY
ATLANTIC OCEAN
RIKERS IS.
RANDALLS IS.
WARDS IS.
GOVERNORS ISLAND
CITY ISLAND
WESTCHESTER & BOSTON LINE
PELHAM LINE
JEROME AVE. LINE
BRONX CONCOURSE LINE
3RD AVE. LINE
QUEENS LINE
FLUSHING LINE
CANARSIE LINE
BROADWAY LINE
FULTON ST. LINE
MYRTLE AVE. LINE
BRIGHTON BEACH LINE
CULVER LINE
SEA BEACH LINE
WEST END LINE
FOURTH AVE. LINE
NOSTRAND AVE. LINE
UTICA AVE. LINE
ROCKAWAY PENINSULA
CONEY ISLAND

b. Design Option B - Greater Neighborhood Transit Access

Figure 30 depicts a new 36-minute, one-seat ride to midtown Manhattan from JFK while restoring the original stops of the Rockaway Beach Branch. Option B will have a newly rebuilt Aqueduct Station as described in Option A.

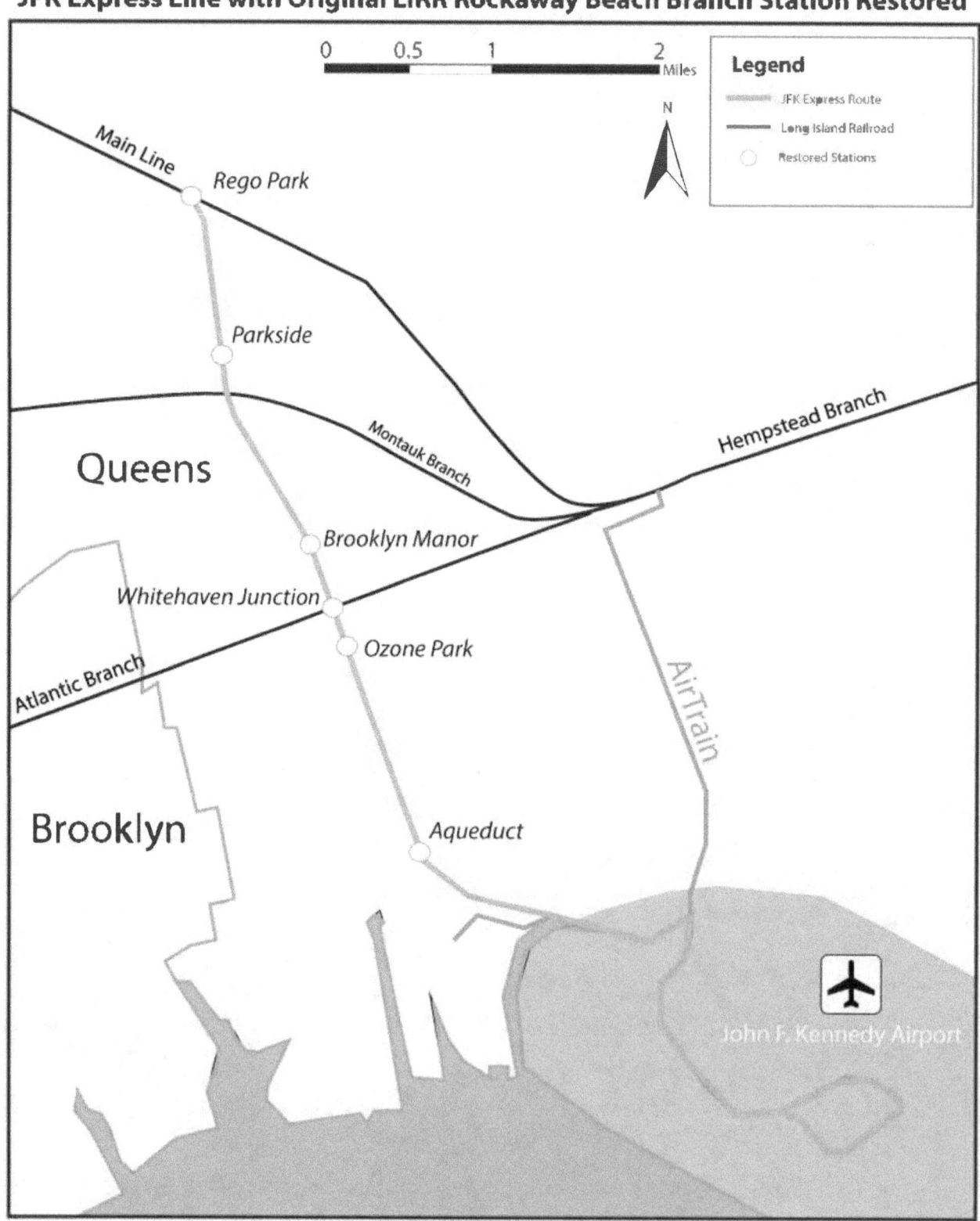

Figure 30
JFK Express Line with Original LIRR Rockaway Beach Branch Station Restored

Figure 31 & 32 are a depiction of a restored local stop. Like Aqueduct, these stations will feature floating canopies over the platforms suspended by steel cables, creating a clean, modern look for the future of Queens rail. The original Rockaway Beach stations will also allow for future connections with the Montauk and Atlantic Branches of the Long Island Railroad.

Figure 31

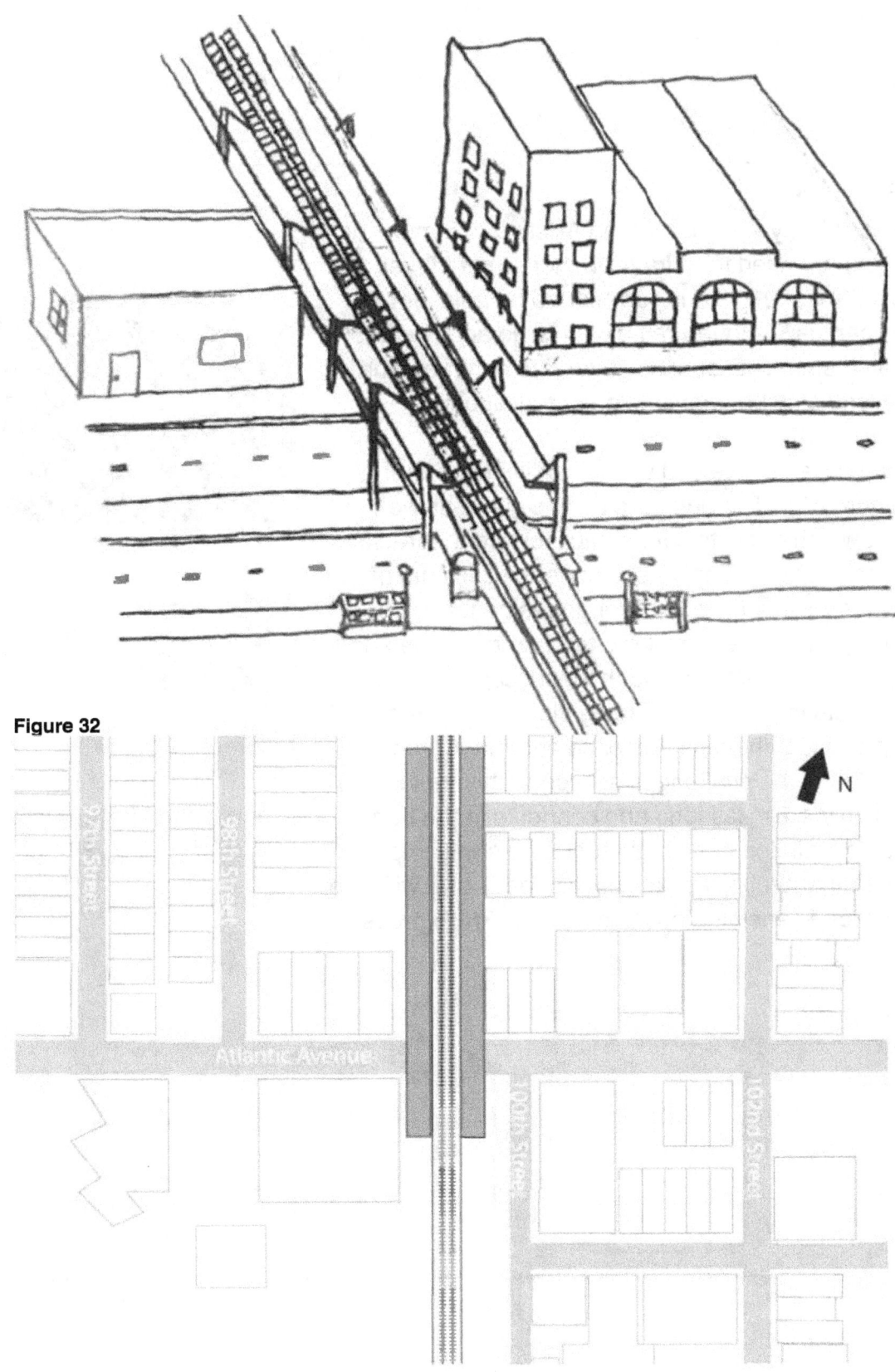

Figure 32

Table 12

Penn Station to JFK - Off-Peak Hours		
Mode	Average Time (Min)	Transfers
JFK Express	36	0
E Train to AirTrain	77	1
LIRR to AirTrain	67	1
A Train to AirTrain	99	1
NYC Airporter Bus	70	0
Taxi	52	0

Table 13

Penn Station to JFK - Peak Hours		
Mode	Average Time (Min)	Transfers
JFK Express	36	0
E Train to AirTrain	66	1
LIRR to AirTrain	46	1
A Train to AirTrain	88	1
NYC Airporter Bus	95	0
Taxi	95	0

d. Sound Barriers

The reactivated railway will run through existing communities that have not experienced train traffic in more than 50 years. Approximately 70% of the line is surrounded by adjacent residential properties.[31] Sound remains a large and legitimate concern. Rail noise mitigations, however, are common and proven successful, having been implemented globally.

The JFK Express was conceived with a cantilever sound barrier (see figure 33). The barrier is built to curve at the top toward the rail, which deflects sound back toward the tracks. This type of barrier is commonly used in dense sections of Hong Kong where active rail runs through residential neighborhoods.[32] The curved design allows the barriers to be shorter and less aesthetically intrusive to the landscape. To further decrease the visual impact, landscaping along the new barriers may help conceal the rails presence and further reduce sound.[33] As noted in the challenges section of this report, the right-of-way is only 4.2 miles long and connects to the Long Island Railroad's Mainline at a sharp curve. This will help prevent trains from traveling at excessive speeds and will curb the amount of sound generated operation of the JFK Express.

'Hong Kong Noise Barrier' Proposal By Architect Francesco Lipari

Figure 33

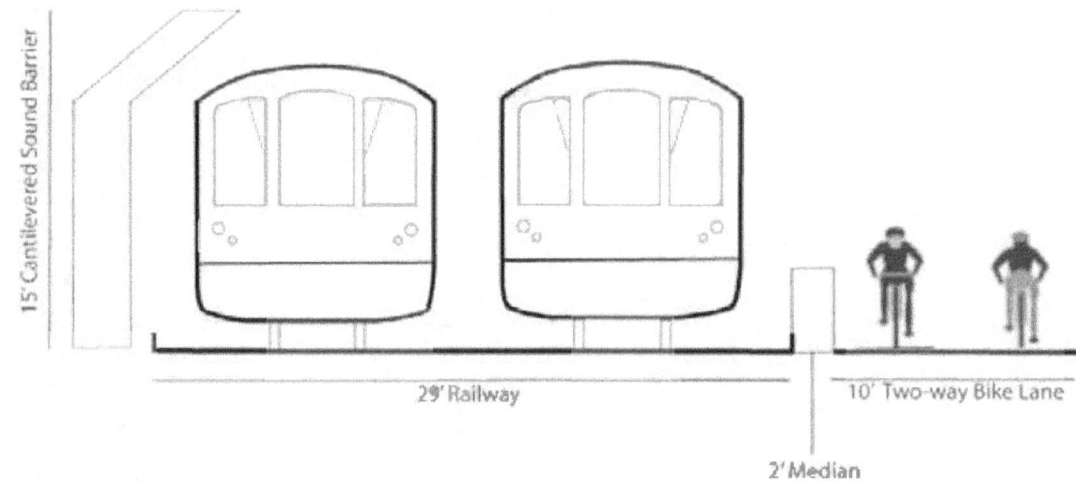

The majority of the disused right-of-way is found on an elevated embankment (see figure 35). As the line passes through Forest Park, the line moves to a sunken embankment (see figure 34). Below is a depiction of the rail with cantilevered barriers for two segments of line.

Figure 34
Sound Barrier on Rail Running within an Embankment

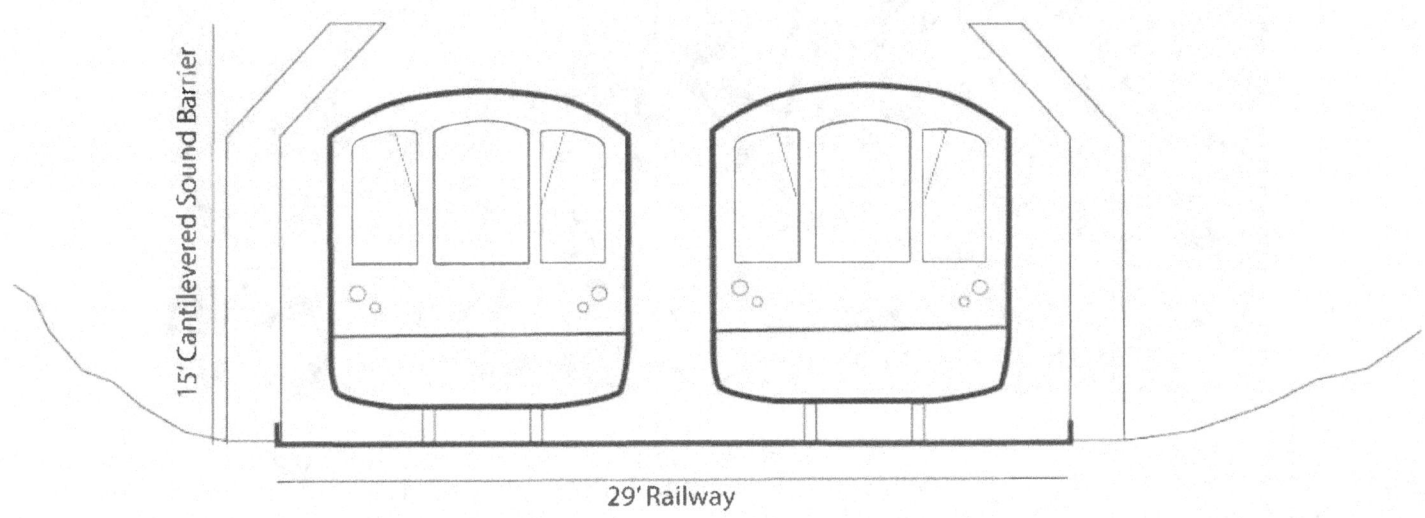

Figure 35
Sound Barrier on Rail Running on an Elevated Embankment

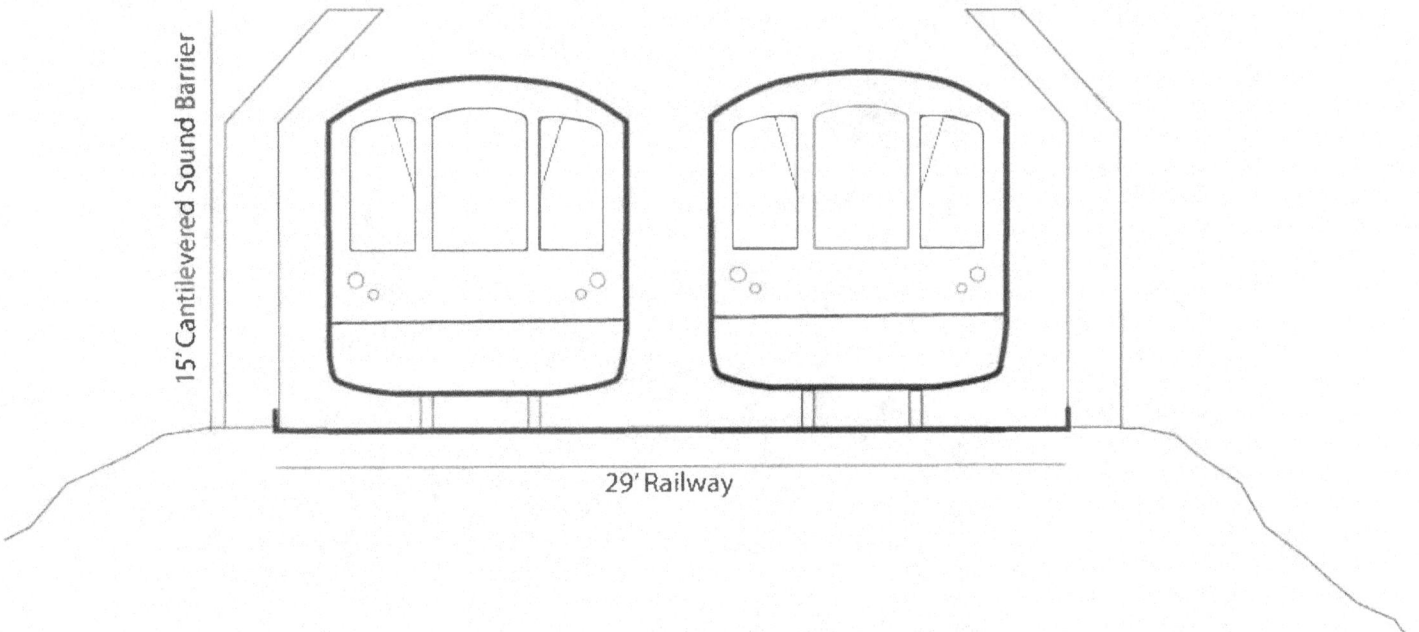

Above the right-of-way's existing conditions in an embankment at the northern section of the disused branch. Residences can be seen to the left with Forest Park to the right. Below is the same image with the proposed design.

e. Incorporation of Pedestrian or Bicycle Path

Stairwells are compact and can occupy sidewalk space at grade crossings.

Rendering from Ross Barney Architects of access for planned Bloomingdale Trail in Chicago. The 3-mile embankment will feature 8 access points from adjacent pocket parks, and a mile and a half of the line will have separated pedestrian and multi-use paths (for bike riders and roller-blades).

One of the greatest characteristics of the Rockaway Rail Branch is the lack of grade crossings as almost all of the line's grade separation remains intact. However, since the right-of-way was designed exclusively for trains, modifications must be made to accommodate pedestrian and/or bicycle uses.

North of Park Lane South, the right-of-way runs at grade and wide enough for two tracks as well as a pedestrian trail or a two-way bicycle lane. Figure 37 depicts approximately 1.5 miles of track which may be ideally suited to accommodate pedestrian or bicycle access (in green). The right-of-way does not appear to be large enough to incorporate both pedestrians and bicycles unless they share a path.

Pedestrian Grade Crossing

Grade crossings for cyclists and pedestrians permit entry and exit points to the right-of-way for pedestrian and/or bicycle paths. Each side of the intersection can feature up to four entry points for these users.

There is only one-separated grade crossing identified for pedestrians at the approximate center of the 1.5 mile pathway. This grade-crossing runs over the Long Island Railroad Montauk freight line (see purple in figure 37).

Figure 36

Cycle entry requires a gradual slope along the transit way.

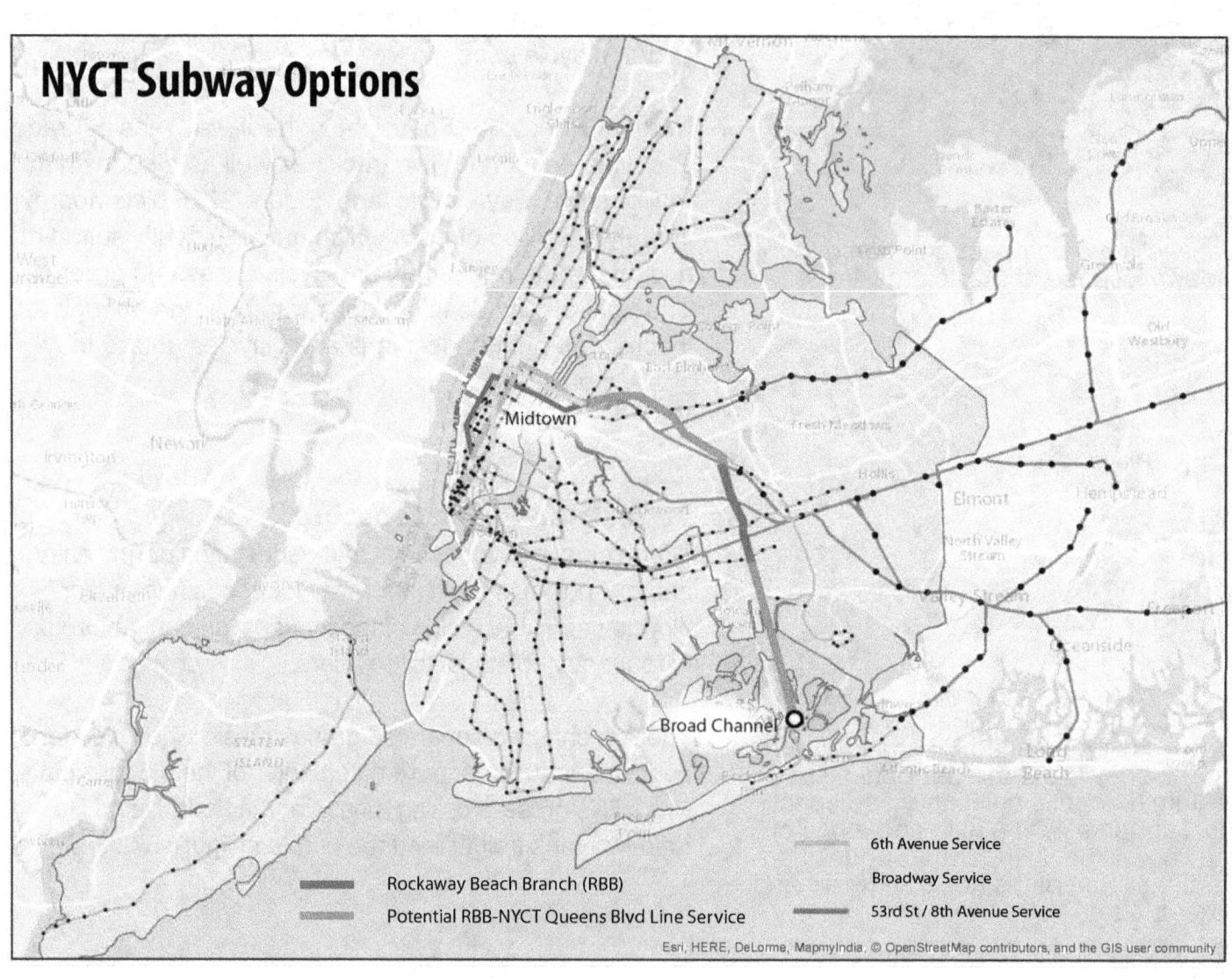

Figure 37
Proposed pedestrian/bicycle greenway along the reactivated rail branch

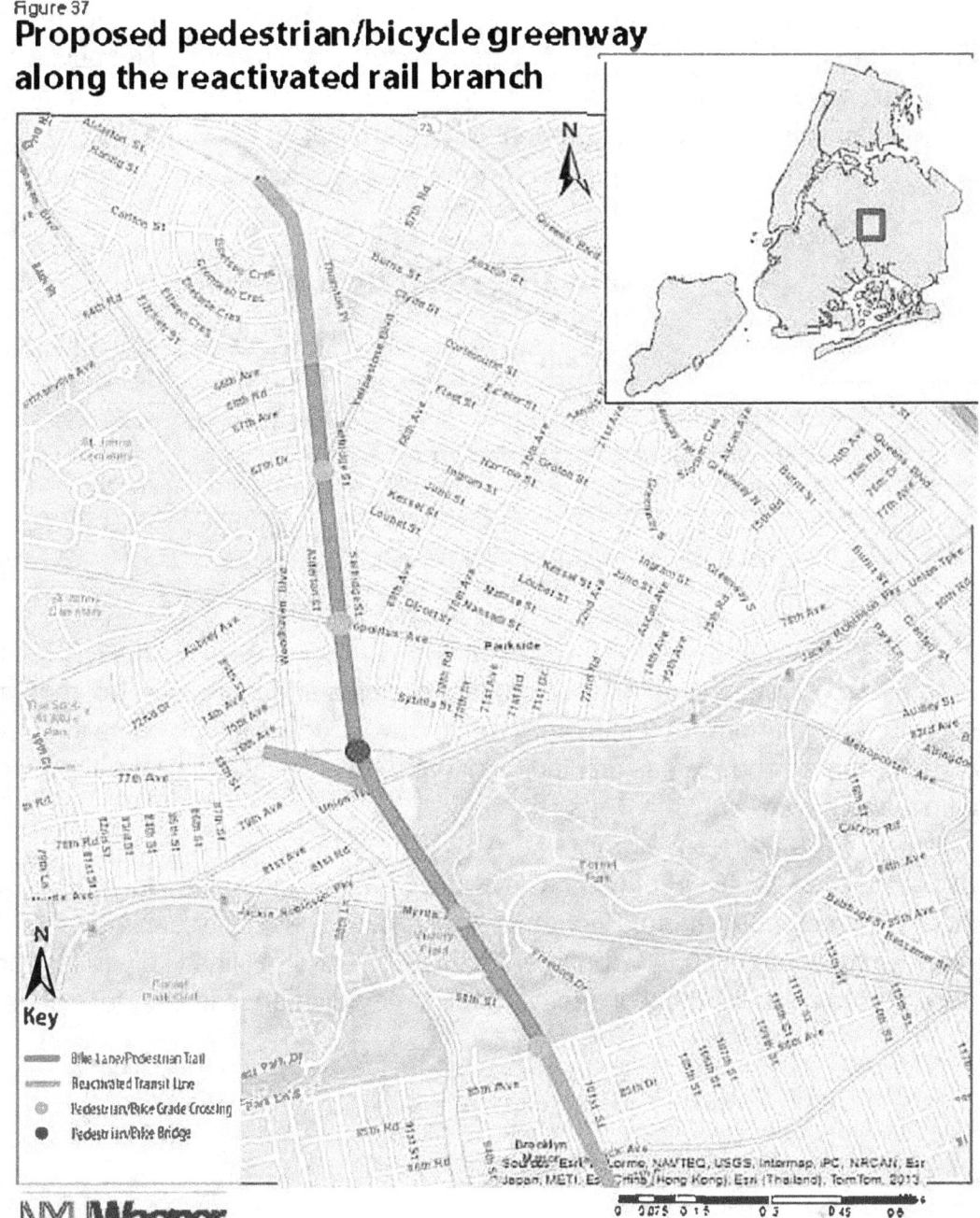

IX. Cost Benefit Analysis of the JFK Express

Costs

Determining an exact cost of reactivation will require a study beyond the scope of this report. Depending on the type of transit implemented prices range drastically. While many studies have provided estimates in cost, the 2001 line study by MTA conducted offers a reasonable basis for comparison. In their study, the MTA projected a cost of $443 million to use the right-of-way to connect with Grand Central Terminal. Adjusting for inflation, the amount to $580 million in 2013 dollars.[34]

Table 14

Reactivation Cost (in millions)

Description	To Grand Central	To Penn Station
Construction	$327	$327
Communications-Based Train Control	$62	$109
Vehicles	$191	$191
Total	**$580**	**$627**

Source: February 2001 'JFK One-Seat Ride Study' - Metropolitan Transportation Authority

Using this cost as a barometer, outside of development benefits and increased tax receipts we expect from this infrastructure investment, we estimate fares to pay off construction costs in as few as nine years. If citywide ridership trends act as any indicator, volume is likely to increase over time, allowing for faster cash flow and payoff.

Taking a macro perspective, the GDP of NYC was about $1.28 trillion dollars in 2010 according the US Department of Commerce Bureau of Economic Analysis.[35] The cost associated with reactivation of the right-of-way represents only 0.04% one year's GDP.[36] For a rail line likely to last decades if not a century or more, this is a small outlay of capital that will pay dividends in its future economic benefit to the city.

> Investing $580 million dollars to construct the JFK Express to Grand Central represents only 0.04% of NYC's 2010 GDP.

Benefits

Increase NYC's Competitiveness in Global Business
As our economy becomes more globalized, businesses require more access to a global network of companies, suppliers and facilities. Having a world-class connection to the airport puts New York City in line with other global finance and economic centers, such as Hong Kong and London, ensuring the city's preeminent position at the top of these sectors.

Decrease Cargo Costs
Reducing private automobile traffic surrounding JFK Airport is paramount to alleviating the sharp decline in cargo volumes at the airport. According to a study by the New York City Economic Development Corporation, congestion of the roadways was a major impediment to the cargo industry in New York. Incentivizing more people to take mass transit to the airport, by offering faster and more convenient service, will help reduce congestion on these roadways.

Greater Transportation Options for Residents in the Rockaways
A free transfer from the 'A' subway train to the JFK Express will greatly access for residents and visitors of the Rockaways. The new line will sharply reduce travel to and from Manhattan.

Job Creation and Growth
True value in reactivating the right-of-way lies in a faster and more efficient commute, but it is far from the only benefit. With reactivation, a variety of new opportunities present themselves to commuters, residents, and businesses.

Construction of the rail will yield a windfall of new jobs both in the short- and long-term. The American Public Transportation Association found that each for billion dollar investment spent, 36,000 jobs are created.[37] Given the estimated costs, reactivation could potentially produce upwards of over 22 thousand jobs with a median salary of $32,100 in May 2011 according to the BLS.[38] A closer look at the top industries of employ suggests that many of these positions can be filled by local workers. The Bureau of Labor Statistics ranks 'Trade, Transportation, and Utilities' among the largest sector by payroll volume, showing that there is a large resource of labor with this skill set.

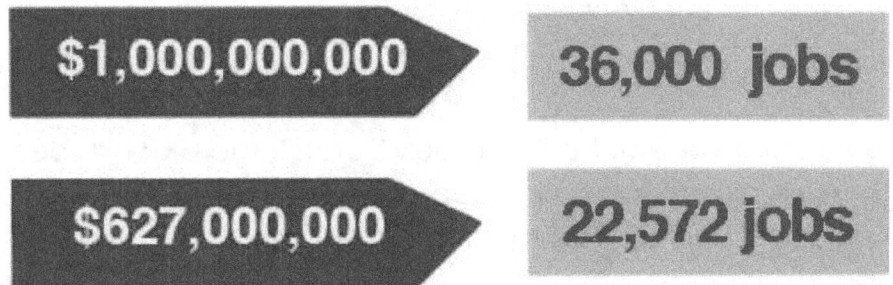

Source: American Public Transportation Association, 2013

Corresponding to the rise in employment will be payroll tax revenues. Given the additional jobs and expected salary, NYC can expect an additional $38.6 million and $23.4 million in city revenues annually.[39] The fare revenue coupled with taxes collected would equal that of the expense of reactivation in just over four years.

BHRA/ ERTRG May, 2017

Some Thoughts on Using Glendale Junction Instead of White Pot Junction
(Lower Montauk vs. Main Line)

Pros:

- Mitigates the political opposition regarding the Forest Park area. See Map 1

- Boosts the "QNS" light rail project, which in turn will lend additional support to RBB.

- The Lower Montauk is only lightly used, and only by freight trains. This means the introduction of a "one seat ride" from midtown Manhattan (Penn Station) to Rockaway would truly be adding new cross Queens rail capacity. This would help alleviate the highly congested, over capacity conditions on the # 7 line, and the Queens Blvd. subway. See map 2.

- The development of the "fourth car" would be justified. First conceived as an integral part of the Second Avenue subway "Manhattan East Side Access" (MESA) plan, this new type of rail vehicle would be able to operate in the subway, emerge to the surface, and then run in City streets as a light rail vehicle. The "fourth car" could also be made compatible with the existing Air Train to JFK. Please see attached NY Daily News article. To read the MESA plan please go to: https://www.scribd.com/document/328995141/MTA-s-Lower-East-Side-Light-Rail-Study-Circa-1999-MESA

- The "fourth car" concept presents a major opportunity to extend NYC subway service into poorly served "transportation deserts" in eastern Queens and western Nassau County – by using less expensive street surface track, rather than highly expensive and time consuming tunneling. Please see Appendix 1 regarding the "76th Street Connection".

Challenges:

- This route relies on the use of the "Montauk Cutoff". The LIRR has been looking to sell or otherwise dispose of this property as a rail to trails, or urban art space. This disposal process would need to immediately cease.

- The connection between the Montauk Cutoff and the portals of the LIRR East River Tubes would require the re-activation of a disused "loop" around Sunnyside yard. This solution is necessary to work around the substantial difference in grade elevation between the Montauk Cutoff and the East River tunnel portals. See Map 2.

- Careful planning would be needed to integrate the QNS/RBB service into the existing train scheduling in the LIRR East River Tubes. Very close coordination would be required from the start of the planning process with the LIRR, Amtrak and NJT.

- The extreme structural decay of the Lower Montauk's bridge over the Dutch Kills would have to be addressed. This ancient and long neglected structure is in need of major repairs, or complete replacement. See attached photos.

NYC DOT

ABOUT DOT [about.shtml]

Lower Montauk Branch Rail Study

Overview

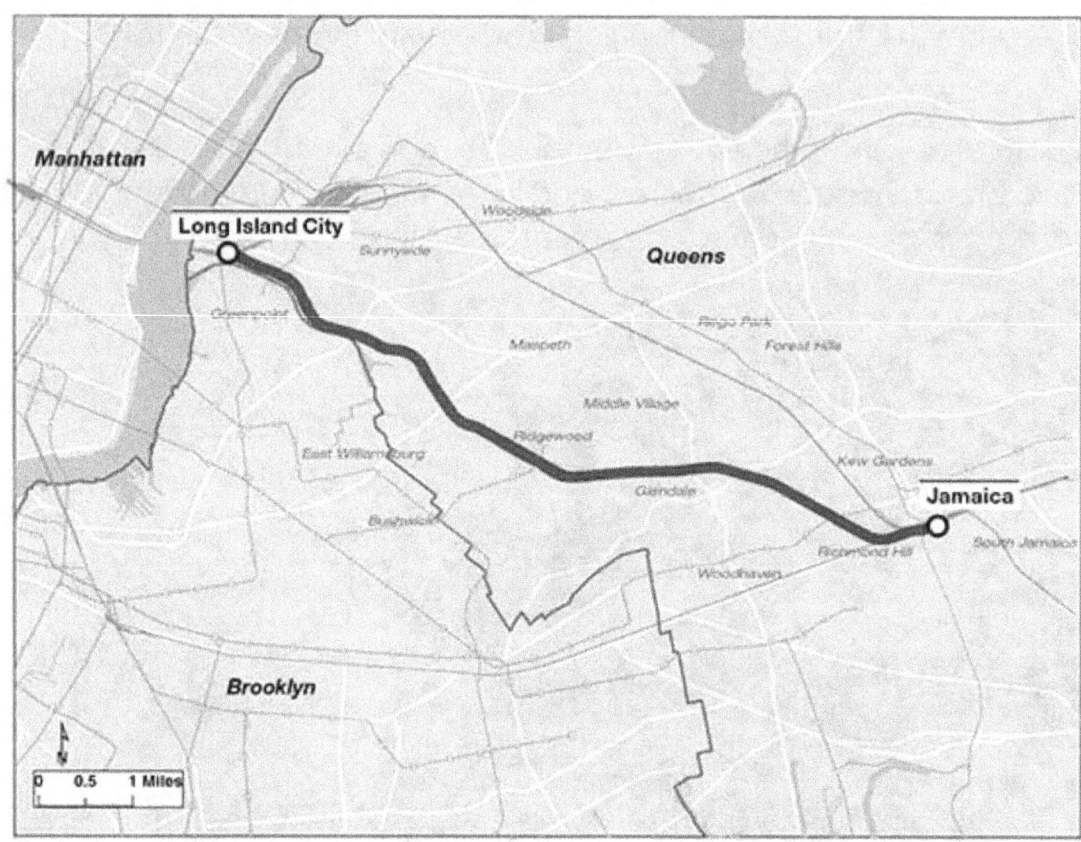

The New York City Department of Transportation is conducting a study to evaluate the feasibility of reintroducing passenger rail service on the Lower Montauk Branch line. The branch connects the existing Long Island Rail Road stations of Long Island City and Jamaica, passing through the neighborhoods of Maspeth, Ridgewood, Glendale, Middle Village, and Richmond Hill. The Long Island Rail Road provided passenger service along the branch at five stations until 1998. Currently, the branch is used for freight service only.

This study, sponsored by Council Member Elizabeth Crowley, will explore options for restoring passenger service that responds to the needs of adjacent neighborhoods, explore development potential to support passenger service, and balance the demands of current and future freight rail activity.

Stakeholder and Public Involvement

May 16, 2017 – Workshop Presentation

>> Download the presentation (pdf) [../../downloads/pdf/lower-montauk-rail-study-may2017.pdf]

January 23, 2017 – Public Stakeholder Outreach Presentation

>> Download the presentation (pdf) [../../downloads/pdf/lower-montauk-rail-study-jan2017.pdf]

Project Timeline

2015– Study Proposed by Council Member Elizabeth Crowley

June 2016 – Funding Secured for Study

Summer 2016– Initial Outreach to Elected Officials, Community Boards, Community Stakeholders and City and State Agencies

December 2016 – Study Initiated

January 2017– Outreach to Elected Officials, Community Boards and Community Stakeholders

Spring 2017– Public Outreach Workshop

Late Summer 2017 – Final Study to be Released

2017-07-19 / Features

DOT Releases Lower Montauk Branch Rail Study

BY THOMAS COGAN

Queens Borough President Melinda Katz and NYC Councilwoman Elizabeth Crowley co-Chaired a July 12 Borough Hall roundtable meeting with local business leaders to solicit their input on Councilwoman Crowley's proposal to reactivate commuter service along the "Lower Montauk Branch" freight rail line that runs from Long Island City to Jamaica. The New York City Department of Transportation (DOT) is currently studying the feasibility of the proposal, which Borough President Katz supports as a means of improving mass transit service in a currently underserved section of western Queens. The DOT is scheduled to complete the feasibility study by the end of December 2017.

The Department of Transportation (DOT) has recently been informing the public about its Lower Montauk Branch Rail Study, which assesses the possibility of reviving a passenger train line that was discontinued almost 20 years ago because of what was thought to be irreversibly declining use.

The latest informational meeting was at Queensborough Hall, where Borough President Melinda Katz and NYC Councilwoman Elizabeth Crowley welcomed an interested audience and turned the slide show over to two DOT executives, Senior Project Manager Jeff Peel and Aaron Suguira, Director of Transit Policy and Planning. But before their description of various proposals for the old passenger line, a brief history of it might be helpful.

Before 1998, the Long Island Rail Road had a branch line known as the Lower Montauk Branch, the last six stations of which were located between Jamaica and Long Island City. On weekday mornings, many riders who were either coming in from Long Island or picking up the train at Jamaica or one of the stations beyond it, would take the train to the LIRR stop in Long Island City, disembark and flock to the No. 7 subway station at Hunters Point Ave., where they added to the passenger load on the train headed to Manhattan, the work destination of most of that subway's riders. The evening rush would produce the opposite effect, with the Flushing-bound No. 7 yielding those many riders at the LIRR switch point.

In 1998, LIRR service from Jamaica to Long Island City was ended, as it was evidently decided that a concentration of ridership during the morning and evening rush hours and nearly none during any other part of weekdays or weekends could be entirely eliminated, leaving those forsaken riders to get to and from Manhattan along the main LIRR line, whose last stop in Queens is Woodside.

Thus, the informational meetings appear to be about reviving the old passenger line, but it's interesting to see that the study's first serious examination is about freight activity in and around the that line. There's a section on the Cross Harbor Freight Project, a vast shipping proposal that has been through years of preparation and controversy seems stronger than ever, given the Port Authority of New York and New Jersey's announcement in May of a Tier II environmental impact study and design funding. The project would stretch from the ports and rail yards of New Jersey across Upper New York Harbor into Brooklyn and then Queens. One result is that the freight rail situation on the Lower Montauk Branch, serious at present, would become far more so in years to come. A bullet point in the study about the "Need to accommodate existing freight and possible growth" is an understatement.

Among what the study calls challenges for freight and passenger operations are space issues. They include the fact that passenger and freight operations need additional yard and track infrastructure; and that there are many right-of-way spots that are quite narrow. Also, everyone has to deal with the feds. Federal Railroad Administration (FRA) safety regulations limit "jointly operating lighter and heavier rail equipment on same or adjacent tracks."

There's also the matter of "crash-worthiness": passenger rail vehicles have to be judged safe to operate jointly with freight equipment, which is much heavier. They might have to operate at different times of the day or never operate on the same tracks and be separated more widely than at present while on adjacent tracks.

Probably the most interesting part of the study for passenger rail advocates is the look at the variety of passenger rail vehicles and what makes them move. The first is (a) the electric multiple unit or EMU, familiar on the LIRR. It needs no locomotive, being operated by third rail power, which would have to be installed on Lower Montauk. It emits little or no pollution and is FRA compliant. (b) Diesel locomotives pulling passenger cars are also familiar. They are FRA compliant and need no electrification, but air and noise pollution issues accompany them. (c) Electric light rail, or "modern streetcar" is self-propelled and powered by a catenary, a cable above the tracks that has a suspended trolley wire. This mode of travel is burdened by what the study calls extensive FRA compliance issues; and of course, the catenary would have to be installed. But like the EMU, it also gives off little or no noise or air pollution. (d) Self-propelled diesel multiple units, DMU, are available too. They have no locomotives and no electrification and the units have what the study calls the availability of FRA compliance options. They are the size of self-propelled electric cars but do have air and noise issues. These are not disadvantageous enough, however, to keep the study from recommending them "for concept development."

In the judgment of the study's authors, lack of electrification wins. To electrify, catenary or third rail systems would have to be constructed; and third rail power, for another thing, requires wider track space, which is precious: it's a third rail after all. The DMU, on the other hand, is ready to go. However, the study admits, though DMUs are less noisy than the huge diesel locomotives and probably pollute less, all diesel trains, large or small, are more sensitive to fuel price fluctuations than electrics.

If only for the purpose of exposition, the study assumes the passenger train plan will proceed. What the study also does is display an assortment of hindrances that come between intention and fulfillment. "Frequent overhead and undergrade crossings may constrain expansion," the study says, for example. On the old rail route there were 11 at-grade crossings, seven overhead and two under grade crossings, and all remain. Widening the separation between freight and passenger trains on adjacent tracks is all but mandated by the FRA, but widening track beds for any reason will be a major procedure, given the rigid confines of elevated viaducts and under grade cuts. At grade level, actual buildings or tunnel walls to either side of tracks will militate against widening the separation between them.

Many other problems are brought up in the study, which boldly says that if all of them can be overcome, ridership on new passenger trains, with restored stations (and possibly new ones), will be ready for service in 2025. For now, mode and station concepts need to be refined; capital and operations cost estimates must be developed; growth scenarios in the train corridor must be projected; and future freight rail activity must be further defined. A later draft report is to be made available before the end of this summer and a final report should appear in the fall.

Return to top

Share / Save

Copyright 1999-2017 The Service Advertising Group, Inc. All rights reserved.

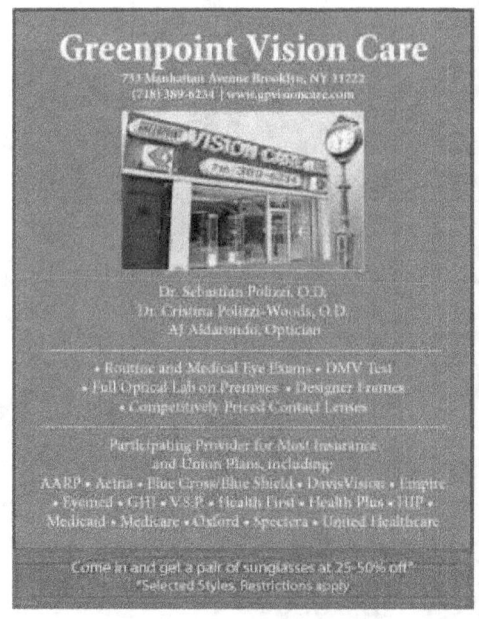

By BHRA / ERTRG, July 23, 2017

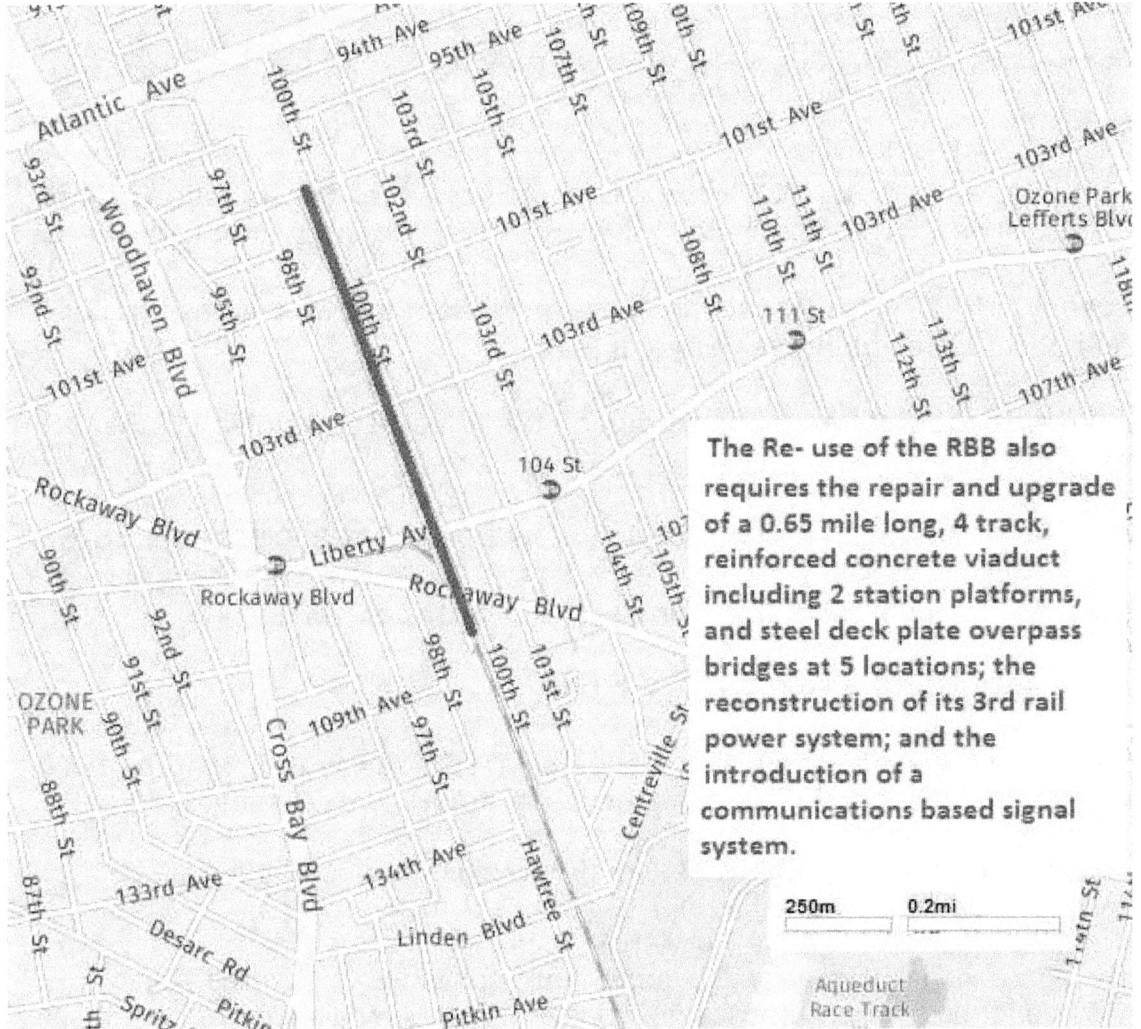

Woodhaven Junction power substation (Atlantic Ave. & Woodhaven Blvd.):

Original primary power was 11kv 25 Hz 3 phase AC (same as IRT subway). Output power was 650 VDC (nominal) to third rail (same as IRT subway).

Type of 3rd rail design on RBB: identical to IRT 3rd rail pickup shoe system. Compatible with BMT "elevated lines" 3rd rail configuration via a "Johnson pickup shoe adaptor cam".

Note: Power substation building still exists. Current photos are available here:
http://www.roadandrailpictures.com/rnywhjct.htm

Construction of the railway section between White Pot Jct. & Glendale Jct. was begun Aug 1908, and completed March 1909. This included the White Pot Junction "fly under". This section was electrified during June, 1910.

Approximate Electric Power Demand

Source: MTA "MESA Study", MES/DEIS, circa August, 1999, pg 12-2, Table 12.1

Notes:

For buses, based on NYCT rates, assumes diesel fuel consumption rate of 0.361 gallons per vehicle mile.

Based on NYCT rates for subways, assumes electric power consumption rate of 5.13 kilowatt hours per vehicle mile.

Conversion factor for fuel consumption by autos/taxis/trucks: 127,700 BTU's per gallon.

Conversion factor for fuel consumption by buses: 131,000 BTUs per gallon.

Conversion factor for electric power: 3,413 BTUs per kwh.

Now, regarding a reactivated RBB:

To be conservative, let's assume a typical NYC subway train length of 600 feet (eight - 75 ft cars). This gives us a "per train" power demand of:

5.13 kwh x 8 cars = Roughly 26 kwh per train (TU) mile. Each train can carry about 1,400 people.

The 4 mile long RBB reactivation route would contain 8 "track miles". Let's assume for estimating purposes, that trains will not be spaced less than ¼ mile apart. These yields a power demand of:

26 kwh/mile x 4miles x 2 tracks = 208 kw
SAY 250 kw x 3 (HVAC, lighting, electronics, etc.) = **750 kw**

HOWEVER, according to NYCT standards, power substations must be able to supply up to 10,000 amps to the 3rd rail "on demand" (the "initial inrush current" of multiple trains).

So, the RBB substation must be able to sustain a peak power output of **6,500 kw (6.5 mw).**

For further detail on the LIRR's original electrification system, please refer to:

https://books.google.com/books?id=0Z9MAAAAYAAJ&pg=PA1010&lpg=PA1010&dq=the+power+transmission+and+third+rail+of+the+long+island+railroad&source=bl&ots=b-becEdfRY&sig=qywJXfyF6Gy6TA9VHeJ7ZwWReNw&hl=en&sa=X&ved=0ahUKEwi3-Laa9JvVAhVMaD4KHcwACIsQ6AEIQTAF#v=onepage&q=the%20power%20transmission%20and%20third%20rail%20of%20the%20long%20island%20railroad&f=false

RBB grade crossing elimination program (extant 4 track earthen embankment, and reinforced concrete viaduct w/plate girder deck bridges over major streets) from Glendale Jct. to Aqueduct was constructed during the 1930's.

Existing "A Train" elevated structure on the Rockaway peninsula was built circa 1939 - 1941.

Source: LIRR Employee Timetable #15 1942, published on June 22, 1941.

Mileage: starting at the original RBB "Beginning Point" (0.0) at "Winfield Jct.":

White Pot (Rego Park)	Old MP 1.7	=	New	MP (0.0)
Parkside	" MP 3.0	"	"	MP (1.3)
Brooklyn Manor	" MP 4.3	"	"	MP (2.6)
Woodhaven Jct.	" MP 4.7	"	"	MP (3.0)
Ozone Park	" MP 5.0	"	"	MP (3.3)
"A Train" Jct.	" N/A	"	"	MP (3.95)
TOTAL				3.95 mi.

SAY 4 Route Miles (8 Track Miles)

NOTE: additionally, the RBB would need to be extended directly into JFK Airport. This would require a new, roughly one (1) mile long viaduct (2 track miles). The precise location, design, and cost, have not yet been determined.

May 8, 1950: Huge Fire destroys LIRR's wooden Jamaica Bay Trestle – LIRR's RBB service ends until the City "buys out" the line.

By 1956, the NYC Transit Authority (pre MTA) completely rebuilt the Jamaica Bay Bridge and began using the RBB line from Liberty Ave. to Rockaway as part of the "A Train" subway line. The City then leased back the RBB line north of Liberty Avenue to the LIRR for its own use. The LIRR then re-abandoned this part of the RBB in 1962. It has lain dormant ever since.

Source: Change at Ozone Park: A History and Description of the Long Island Rail Road Rockaway Branches, by Herbert George, 1993

IND Queens Line 33rd Drive Station: during the circa 1930's IND Queens Blvd. subway construction, provision was made for direct connection to the RBB. This Connection Provision Still EXISTS. Please read:

https://en.wikipedia.org/wiki/63rd_Drive%E2%80%93Rego_Park_(IND_Queens_Boulevard_Line)#Unfinished_Rockaway_spur

A direct extension of the RBB into JFK AIRPORT was contemplated back in the 1970's. Please read the following extract from the THE PORT AUTHORITY OF NEW YORK AND NEW JERSEY JFK INTERNATIONAL AIRPORT LIGHT RAIL SYSTEM, Final Environmental Impact Statement, May 1976, Vol 3, PDF page 872

This Space Left Intentionally Blank

October 10, 1996

Torin Reid
120-27 Marsden St.
Jamaica, NY, 11434

Mr. Lawrence Schaefer
FAA, AEA-620
Fitzgerald Fed. Bldg. JFK Airport
Jamaica, NY, 11430

Dear Mr. Schaefer,

    I apologize for the lateness of my letter. As for my reply to the FAA Written Revaluation/Technical Report, I submit to you a condensed manuscript of mine, that proposes using the LIRR abandoned right of way in Queens. I believe that this is worth your consideration.

Sincerely,

Torin ..eid

THE RAILWAY TO THE AIRWAYS

PART ONE: THE PAST

I'm quite sure more people than simply the area railroad enthusiasts are aware of the old, abandoned railroad that runs through southeast Queens. Plenty of people who live adjacent to it are knowledge-able. Some politicians know about it. I'm sure that even the Port Authority (of NY & NJ) know about it.

But I would like to tell you a little more about this old railroad, and how it could serve a new and useful purpose today as a line to JFK Airport.

This old railroad was originally built as the New York, Woodhaven and Rockaway railroad, constructed in 1877. It connected with the Long Island railroad at what was called Glendale Junction, on the present day LIRR Montauk branch near Woodhaven Boulevard. The NY,W&R built south, connecting again with the Long Island's Atlantic branch between today's Atlantic and 95th Avenues. The line continued straight across South Ozone Park, down to and across Jamaica Bay, on a long wooden trestle, to the Rockaway Peninsula. There, it turned due west, and ended at Rockaway Park at the site of the present subway terminal there. At that time, the Rockaways were a vacation and resort area, and the New York, Woodhaven and Rockaway sort of pre-empted the Long Island - which had built an earlier branch over a longer, all-land route via Valley Stream and Far Rockaway.

After ten years of cooperative operation, the Long Island took over the New York, Woodhaven and Rockaway in 1887. The next year, the Long Island built a connection from the NY,W&R line near the entrance to the peninsula and connected it to the earlier Far Rockaway branch.

This formed, in railroad parlance, a "wye" which was somewhat south of today's Hammels Wye where the subway line splits to go to Far Rockaway and Rockaway Park.

Over the years, the Long Island railroad rebuilt the line, extending it northward from Glendale Junction towards the main line in Rego Park. The line was elevated on an embankment, and after awhile, on a concrete viaduct in South Ozone Park. Later still, the lines in the Rockaways were elevated above street level. Still, the Achilles' heel of the line was that wooden trestle across Jamaica Bay, which had to be rebuilt time and again after fires. In 1950, a weakened, bankrupt LIRR, reeling from two disastrous wrecks that year, could no longer afford to rebuild the trestle and so abandoned that part of the railroad.

While the Long Island continued to operate trains to Rockaway park via Valley Stream, the railroad wished to drop all service on the Rockaways. So, in 1952, the whole Rockaway line was sold to the New York City Transit System. A branch line was built down from the old Liberty Avenue elevated line (today's A to Lefferts Blvd.) which crossed the Long Island in South Ozone Park. The Long Island continued to operate the line as a spur north of South Ozone Park until 1962. The New York City Transit System completely filled in the old trestle across Jamaica Bay, added two swing bridges (only one is in use today) and servered the line at Mott Avenue in Far Rockaway. The Long Island retreated back across the Nassau County border and built it's own Far Rockaway station.

With the building and expansion of nearby airport, no thought was given to the railroad as a means of access.

THE RAILWAY TO THE AIRWAYS
PART TWO: A POSSIBLE FUTURE

Fast Forward to 1996. Kennedy Airport has a worldwide reputation for being a intimidating, frightening introduction to New York. One of the biggest complaints from airline passengers is that cab ride to Manhattan. What can we do?

This is where the old New York, Woodhaven and Rockaway railroad come in. Since my first trip in 1990, I have walked the abandoned line several times. I am happy to report that time and nature has been kind.

From the Long Island railroad mainline near 63rd Drive in Rego Park, the sounds of the LIRR trains dies down as I tromp into the thickening underbrush. On the ground below, rails and railroad ties are shrouded by undergrowth - one must step carefully. At first, the tracks are widely separated - what was the westbound, or Manhattan-bound track dived down and under the mainline in a tunnels still extant. The tracks come together on a high embankment, bridging only the major streets in Queens. Although I am no civil engineer, many of the bridges look to be in fairly good shape, and some have been painted in the thirty four years since the last train rolled through. Thousands of trees have grown in the right of way, but paths have been worn through by exploring neighborhood children. It is often still, and quiet here, except when crossing the streets. At the site of the old Glendale Junction, where a burned out wooden trestle once carried the line over the LIRR Montauk branch, a connecting track can be made out in the brush behind a small baseball diamond. One of two man made disturbances occurs near here, where a housing developement has taken over the right of way for several hundred yards as parking space for it's tenants. Often, there are views of many backyards - which makes me think that this railroad would be a far more pleasant introduction to New York than a sneering driver on the Van Wyck expressway.

There were several stations on the line, but only two still exist - Woodhaven Junction (at Atlantic Avenue) and South Ozone Park, near 101st Avenue. At Woodhaven Junction, a power substation remains as a shell, but there is little evidence of the connection to the underground LIRR Atlantic Avenue line to Flatbush Avenue, Brooklyn. Further south, a school bus operator has bulldozed the line to make room for his buses.

On the viaduct through South Ozone Park, it is easy to trace the switches and crossovers as the railroad widens to four tracks. The line continues until obliterated by the TA just south of Liberty Avenue. Here, one can see planes preparing to land at Kennedy.

The best part is this: south of Liberty Avenue, the right of way remains at the width of four tracks. The TA uses two tracks for the A line. By moving over the Manhattan bound A line track next to the Rockaways bound A line, we make space for another two track line - to the airport. The subway track needs to be moved only between Liberty Avenue and Howard Beach, where the subway is closest to the airport. In addition, there is a downgrade between the station at Conduit Avenue and Howard Beach. When we keep the two airport line tracks at the higher level, we create a viaduct or trestle, which then curve eastward across the site of the PA parking lot at Howard Beach, to the airport proper, which is about a mile away at this point. This would be a true airport line that would elevate New York to that world class of cities that have easy airport access, like London, Frankfurt, Chicago and Cleveland (Cleveland? Don't laugh, that was the first airport line in the U.S.).

The stations at Aqueduct and Conduit Avenue would have to go, although Conduit Avenue could be rebuilt as an island platform (between two tracks).

15a | But, think about it - the greatest expense would be building that viaduct a mile or so from Howard Beach to the airport - and reconstructing the right of way to Rego Park in Queens. Plugging this line into the Long Island railroad gives you instant access to Penn Station. How easy can it get?

One more thing - How long does it take to get there? Well, a look at some old LIRR timetables reveals that trains made it from Penn Station to South Ozone Park station in 23 to 25 minutes (!). I would add a very conservative 15 minutes more to get to Kennedy Airport. All told, we're talking 40 minutes from midtown Manhattan to Kennedy. Is this a winner, or what?

A SOLUTION FOR LaGUARDIA, TOO

Unfortunately, La Guardia Airport is a place where an arriving passenger seeking to get into the city is confronted with a plethora of taxis, black cars, and hustlers with station wagons. Once again, this can be an intimidating prospect, even for city residents. The traveller is not even likely to see a Triboro Coach Q-33 bus or a MTA Q-48 bus; these are lost in traffic somewhere.

I suggest that the Port Authority to build an AGT line, similar to the one abuilding at Newark, to serve LaGuardia. It should run, not to some packed side street in Manhattan, and not to Kennedy airport, but it should run to the Willets Point - Shea Stadium station of the No. 7 line. This station is a natural, not only because of it's wide walkways for the ballgame crowd (although they were rebuilt for the 1964-65 World's Fair), but also because the LIRR Port Washington branch is but a few hundred yards away, an alternate route to Manhattan!

The route for this AGT would be along the east side of the Grand Central Parkway, then it would curve along the subway line next to the Roosevelt Avenue viaduct. The AGT must, however, fly high over the ganglia of roads that connect the Grand Central, Northren Blvd., and Van Wyck expressway at one point.

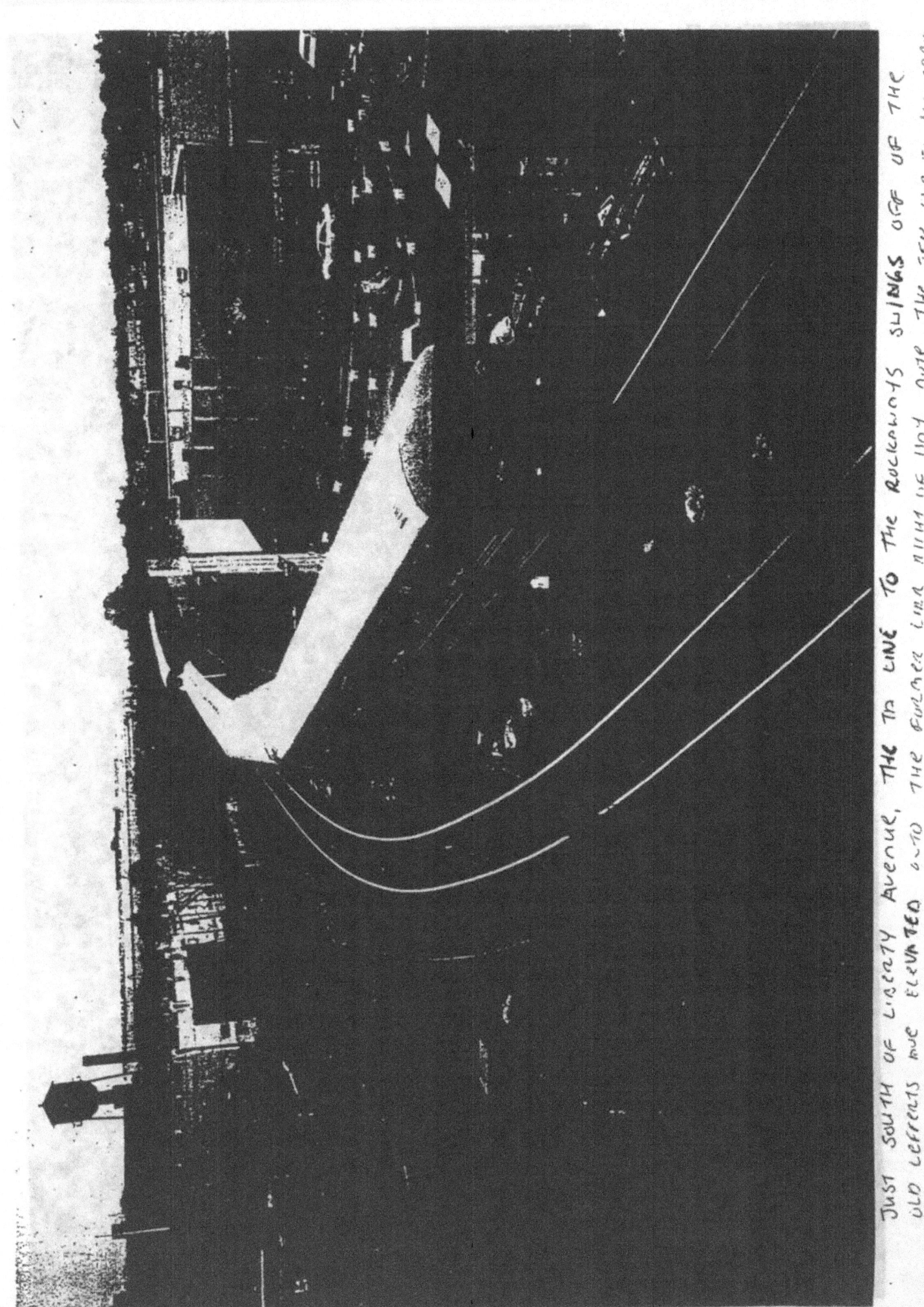

JUST SOUTH OF LIBERTY AVENUE, THE TO LINE TO THE ROCKAWAYS SWINGS OFF OF THE OLD LEFFERTS AVE ELEVATED & TO THE FULTON LINE NIGHT BE USED TO THE END TO HUDSON...

THE OLD TRACKAGE EXTENDS RIGHT UP TO THE PRESENT LINE RIGHT OF WAY IN FOREST HILLS

"The Fourth Car"

New Type of Light Rail Vehicle That Can Operate in NYC Subways, on the LIRR, on the AirTrain, and on City Streets

---

A One Seat Ride from midtown Manhattan to JFK Airport, via the QNS Light Rail and a Re-Built LIRR Rockaway Beach Line

May 16, 2017

Vehicle Design Requirements and Characteristics

The following section describes the design requirements and characteristics for a rail passenger vehicle that could be feasibly constructed and operated on both the LIRR and AirTrain LRS systems.

General Arrangement

One-Seat Ride vehicles would be arranged in a 4-car fixed unit with a maximum length of 240 feet. This would ensure that One-Seat Ride trains do not exceed the length of LRS station platforms. A 4-car fixed unit is preferred because (1) the demand for service during most of the service day would require at least the 152-seat capacity of these units and (2) economies of scale associated with the maintenance of a fixed unit compared to individual cars. Each unit would include two types of cars, designated as "A" and "B." The cars at the outer ends of each unit would be "A" cars; an operator's cab would be provided at one end of each "A" car. Each cab end would be equipped with other features such as energy absorbing zones necessary to meet the LRS and LIRR crashworthiness requirements. The intermediate cars would be designated as "B" cars.

The spacing of the automated sliding doors along LRS station platforms would govern the door arrangement for the One-Seat Ride unit. As a result, there would be two doorways per side in each car, or eight on each side of each One-Seat Ride unit.

A full-width operator's cab would be provided at each end of the unit, to be used for the portion of the trip on the LIRR. To provide a seamless transition to and from the LIRR's operating environment, the cab and its controls would resemble that of the LIRR's M-7 cars as closely as possible in appearance and arrangement.

Exhibit 24 summarizes the overall dimensions for a One-Seat Ride unit composed of empty cars ready for revenue service, without crew or passengers. Exhibit 25 provides a schematic of a One-Seat Ride "A" and "B" car.

Exhibit 24
One-Seat Ride Vehicle General Dimensions

Vehicle Element	Dimension
Unit length over coupler faces, maximum	240 feet, 0 inches
Width, overall maximum	10 feet, 6 inches
Height, nominal, top of rail to side door thresholds, LRS position	3 feet, 8 inches
Height, nominal, top of rail to side door thresholds, LIRR position	4 feet, 2 inches
Height, side door thresholds to top of roof	8 feet, 7 inches
Side door clear opening, minimum height	6 feet, 3 inches
Side door clear opening, minimum width	5 feet, 0 inches
End door clear opening, minimum height	6 feet, 3 inches
End door clear opening, minimum width	2 feet, 6 inches
Maximum truck wheelbase	7 feet, 0 inches
Wheel diameter, new, minimum	28 inches
Passenger capacity	152 seated/236 maximum

Exhibit 25
One-Seat Ride Vehicle Schematic

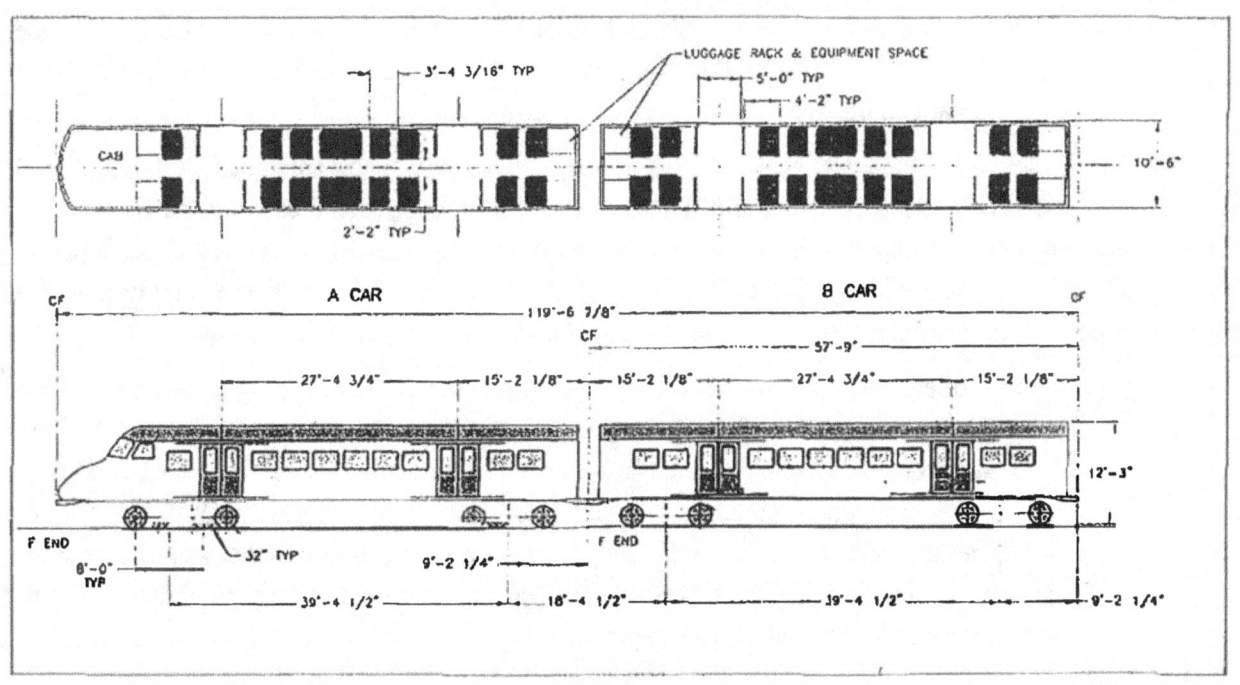

Service Quality

The One-Seat Ride would be offered as a fast, frequent service targeted to airline passengers. As a result, the One-Seat Ride vehicle would need to incorporate the following features:

- **Industrial Design and Styling.** The One-Seat Ride vehicles would need to present a pleasing, ultra-modern appearance, inside and out. The vehicle specification would require the involvement of an approved industrial designer throughout vehicle design and construction, and the provision of renderings and mockups for review and approval.

- **Car Layout.** The One-Seat Ride vehicle should offer a high level of comfort. Therefore, the vehicle would be designed to provide seats for practically all passengers. Seats would have generous dimensions, with two-passenger seats on each side of the aisle. Because of the need for a quick turn-around in Manhattan, and to avoid labor costs for turning seats and maintenance problems, seats would be fixed, half facing in each direction.

    Luggage rack areas would be provided in each car. Since the new FRA crashworthiness standards and the voluntary Passenger Rail Equipment Safety Standards developed by the American Public Transportation Association both require a crush zone at each end of each new passenger rail car, the luggage racks would be located in this area to avoid wasted space.

- **Ride Quality and Noise.** The One-Seat Ride vehicles would be designed to be free from objectionable vibration and shock. All equipment mounted in the passenger area would be free from resonance to avoid annoying audio and visual distraction. Additionally, noise levels under all conditions would be no greater than those generated by the LRS vehicles or by the LIRR M-7 cars under similar conditions.

- **Amenities.** The One-Seat Ride vehicles would feature passenger amenities designed for air travelers such as ample seating and luggage racks. Additional amenities could be incorporated into the design of the One-Seat Ride vehicles based on the policies and standards established for the service. The following describes two amenities that could be considered for the One-Seat Ride vehicles as well as their associated technical and institutional considerations:

    - **Toilets.** Toilet rooms are not provided in the One-Seat Ride vehicles because each would displace six to nine seats. Toilet rooms would also add significant weight, which would need to be evaluated relative to a One-Seat Ride train's fully loaded weight and the limits of the LRS' elevated structures. Toilets would also require servicing, which cannot be done on the loop at the airport, and would be difficult at the Manhattan terminal. Moving the unit to a point where such servicing can be done may ultimately require an additional unit to cover the service requirement, at considerable expense.

- **Real-Time Passenger Information.** A real-time video display of departing flights from the airport would be a convenience for JFK Airport-bound trips. With multiple terminals and dozens of airlines involved, assembling all arrival and departure information in real time may be an institutional challenge. If flight information can be gathered comprehensively and reliably, transmission to the train and on-board video displays would not be particularly difficult.

AirTrain LRS and LIRR Requirements

The One-Seat Ride vehicle must meet the physical, operating, and regulatory requirements of both the LRS and the LIRR. The consultant team has identified strategies for addressing differences between the two systems. A compromise design could accommodate some of these differences, and resetting system parameters as the vehicle moves from one system to the other could accommodate some other differences. A few would require separate equipment for use on each system. The following describes how AirTrain LRS and LIRR requirements would be addressed within the design of the One-Seat Ride vehicle:

- **FRA Requirements and Weights.** FRA has recently issued new rules governing the construction of passenger cars for use on railroads under its jurisdiction, such as the LIRR. The rules for locomotives have also been revised recently. These new rules set requirements for structural strength and crashworthiness, which has the effect of increasing vehicle weights. The challenge for the One-Seat Ride vehicle is to achieve FRA's standards while not exceeding the weight limits of the LRS' structures. Achieving this goal will require careful design and avoidance of any unnecessary weight. Exhibit 26 shows the approximate weights for the LIRR, LRS, and One-Seat Ride vehicles and weight limits for the LRS.

Exhibit 26
One-Seat Ride Vehicle Weight Parameters

System Element	Weight Status	Weight
LIRR M-7 cars, projected weight	Empty	60+ tons
Above, adjusted to 58-foot car	Empty	47+ tons
Permitted weight in 58-foot car on the LRS	Crush Load	46 tons
Passenger load in One-Seat Ride/LIRR vehicle	Crush Load	9 tons
Maximum weight for One-Seat Ride/LIRR vehicle	Empty	37 tons

While it is intended that the One-Seat Ride service would not normally carry the maximum load of 236 seated and standing passengers, the standing area of a train may be filled to a crush load during a possible service disruption. Deducting the weight of the crush load from the allowable gross weight leaves little weight for the vehicle itself. This is addressed in the conceptual vehicle design by providing enough seats that are positioned to minimize the standing area and thus the possible crush load.

- **Crashworthiness.** The LRS and LIRR signal systems are designed as the primary protection against collisions. Both systems' rail vehicles are also designed to meet certain crashworthiness standards to limit the risk to passengers in the unlikely event of a collision. The LRS system is being designed as a stand-alone system with crashworthiness requirements appropriate for the weight and strength of the LRS vehicles. The One-Seat Ride vehicle would have to meet FRA crashworthiness standards and be strong enough to protect its passengers in a collision with LIRR equipment. At the same time, the One-Seat Ride vehicle must not excessively damage an LRS vehicle if a collision should occur on the LRS system. To meet these requirements, the One-Seat Ride vehicles would have both energy absorbers and an outer crush zone at each end of the unit. These would lessen the shock of a collision.

- **Coupler and Platform Height.** On commuter railroad cars such as those operated on the LIRR, the coupler of one car is retained under the end sill of the other car to keep the heavy underframe structures of colliding vehicles in line. This maximizes resistance to the penetration of one vehicle by another. However, transit vehicles such as those on the LRS use an anticlimber, a horizontal member across the end of the car with ribs that engage those of the other car to resist vertical movement. To meet FRA requirements, the One-Seat Ride vehicle would have a coupler height in line with, not below, the anticlimber of the LRS vehicle. To address this issue, the One-Seat Ride vehicle would have both a transit anticlimber and a coupler meeting FRA strength requirements. The height of the One-Seat Ride vehicle body would adjust above the track as the vehicle shifts between the LRS and the LIRR. This would properly align the anticlimbers when the One-Seat Ride train operates on the LRS, and the couplers when operating on the LIRR. For the recommended alternative, L2, this adjustment would take place at the AirTrain station in Jamaica.

The adjustable car body also would address differences in platform height between the LIRR and the LRS. The mechanism would provide a vertical movement of about six to seven inches to meet the platform and coupler/anti-climber height requirements of both the LRS and the LIRR. A similar arrangement of a passenger vehicle capable of elevating to different heights is found in the mobile lounges used at several airports, including Montreal-Mirabel, Dulles International (Washington, D.C.) and at JFK Airport, for certain flights unable to use terminal jetways. One of the firms that built these mobile lounge vehicles was the Budd Company (since acquired by Bombardier), which also built the LIRR/Metro-North M-1 and M-3 fleet and NYCT's R-32 cars.

- **Mode Selection.** Because of the differences in train control, communications systems, and car body height required on the LRS and the LIRR, the One-Seat Ride unit would be equipped with a mode control to adapt it to LRS and LIRR requirements. When the LRS mode is selected, the affected systems used on the LRS would be activated, and the car body would be moved to the lower position. When the LIRR mode is selected, the systems used on the LIRR would be activated and the car body would be moved to the upper position.

- **Communications Systems.** The LRS and the LIRR use different communication systems for a variety of purposes, including train radio, public address, passenger information, intercom between cabs, and diagnostic data exchange with wayside installations. As a result, the One-Seat Ride vehicle would feature a complete communications system that would provide all required functions for operation on the LRS and the LIRR. Where the functions necessary for operation on one rail system are different from those required for operation on the other, separate functional units would be provided.

- **Cab Layout and Facilities.** To meet LIRR operating requirements, the One-Seat Ride vehicle cab would be outfitted with manual operation equipment that would allow the trains to be operated on LIRR tracks between Manhattan and the connection point to the LRS system. Equipment and control layout in the One-Seat Ride cab should be as similar as possible to that of the M-7 car. On the LRS, One-Seat Ride trains would be operated in an automated mode similar to that of the on-airport trains. As a result, the One-Seat Ride vehicle would need to include LRS-compatible automated train control equipment as well.

- **Maintenance Facilities.** The One-Seat Ride vehicle specification would require that the vehicles be as compatible as possible with both systems' maintenance facilities. Even when one system has been selected to perform normal maintenance, it would always be possible that, in case of a breakdown, the most expeditious course would be to take the unit to the nearest facility.

Vehicle Costs

The per-unit vehicle cost for alternatives L1 and L2, either of which would require 32 cars (including spares) to maintain a 15-minute headway, is estimated at $3.96 million, for a total cost of $127 million in 1999 dollars. For alternatives L8 and L9, either of which would require 40 cars, the estimated per unit vehicle cost is $3.65 million, for a total cost of $146 million in 1999 dollars. These vehicle cost estimates are based on as assessment of industry standard and unique features of the One-Seat Ride vehicle, the size of the fleet order, and the LIRR's recent fleet purchase experience. The costs reflect the fact that to operate on both the LIRR and LRS systems, the One-Seat Ride vehicles require a custom design. The costs include activities associated with engineering, manufacturing, administration, and testing.

Rockaway Park Branch — 63rd Drive to Fleet St.
elf 11.58

Main Line #3
" " #1
" " #2
" " #4
Elec
"
"
"

#1 Elec
#2 Elec

63rd Drive

63rd

for Cont. wire M.Line

Sig. Br.

MP 6

Motawok Station
New 1923
out ?

L. Plat.

#1 Elec
#2 Elec (only)
out (1929) — 15
R63d on
Jct 7.57
Dk
6.55
55 ST
Late AGS 62
1908

Fleet St.
(ex Washington Rd.)

1. New S.S. & T — in 10:58
2. " — End of auto Block Sig
3. " " Manuel —
4. Ex Sipe T — out 10:58
5. Ex Sand Pit Track, Track Removed 1928
6. To 6 — out 10:58
6. To " — in 10:58
77. Ex S.T. of Matawok Land Co. — overland towards w 6.S. 5.5 (Private) in N77 out?

19 — Stairs to footbridge

eff. 11:58 FLeeT ST. To METROPOLITAN Ave Rockaway PARK BR.

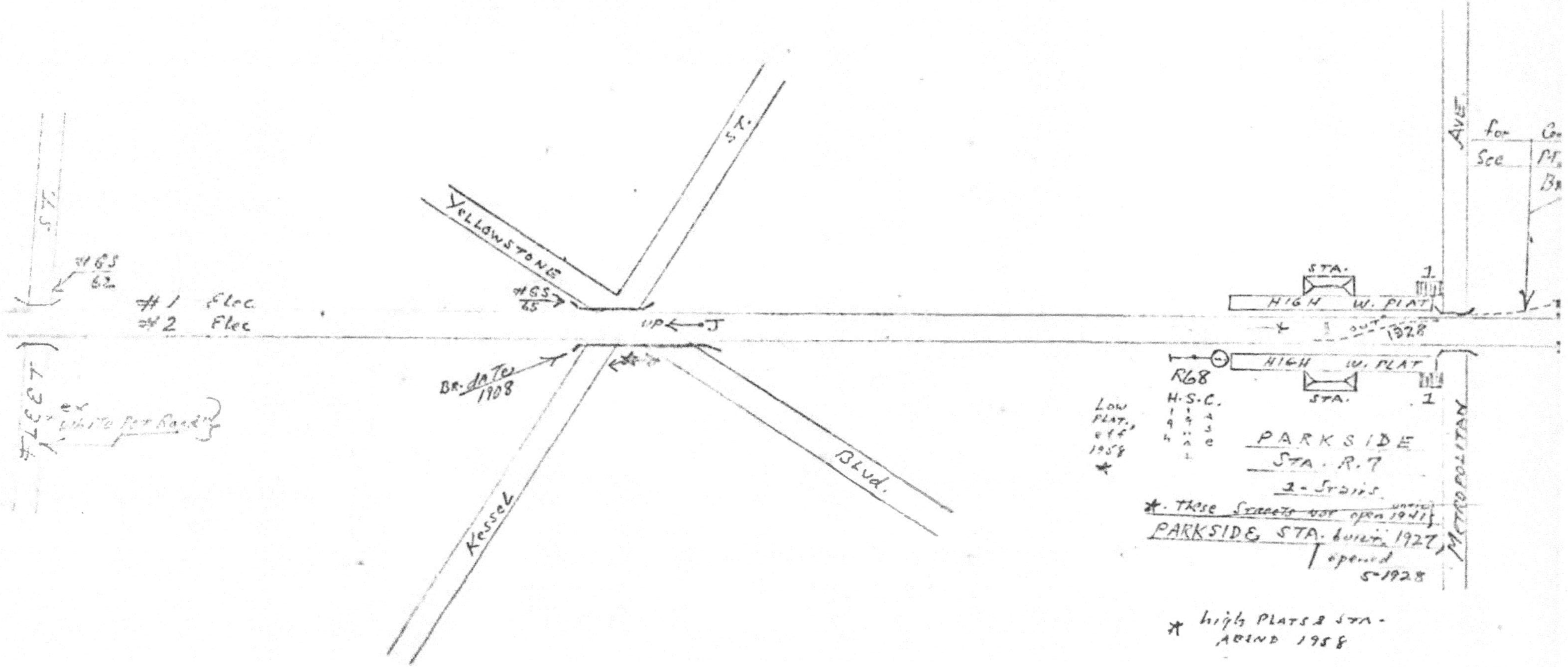

FLEET ST.
#GS 62
#1 Elec
#2 Elec
ex White pot Road
YELLOWSTONE
ST.
#GS 65
UP J
BR. date 1908
Kessel
Blvd.
STA.
HIGH W. PLAT
HIGH W. PLAT
STA.
OUT 1928
R68
H.S.C.
Low PLAT. eff 1958
PARKSIDE
STA. R.7
2 - Stairs
*. These Streets not open until 1941
PARKSIDE STA. built 1927
opened 5-1928
METROPOLITAN
AVE
for See
* high PLATS & STA. ABAND 1958

Rockaway Park Br.   eff 11:58   MYRTLE AVE. TO 91ST. AVE.

BROOKLYN MANOR ★
STA. R 73/4
★ high platf. glass
above 1958

★ Bkln Manor Sta. opened 1911, replaced
Brooklyn Hills Sta. at ⓐ, which closed
in 1911. (Low platforms at ⓐ only)
For Trolley Hwy To 91st Ave see Map R2
" Jamaica Ave " " " " 91st Ave " " R4

1. Water Tanks of Glendale Wells?
2. Ground Level Coal Pockets
▲ 3rd Rail Removed 1959
3. Stairs to St.
"A to A" Ex R&Q T Trolley Line Route

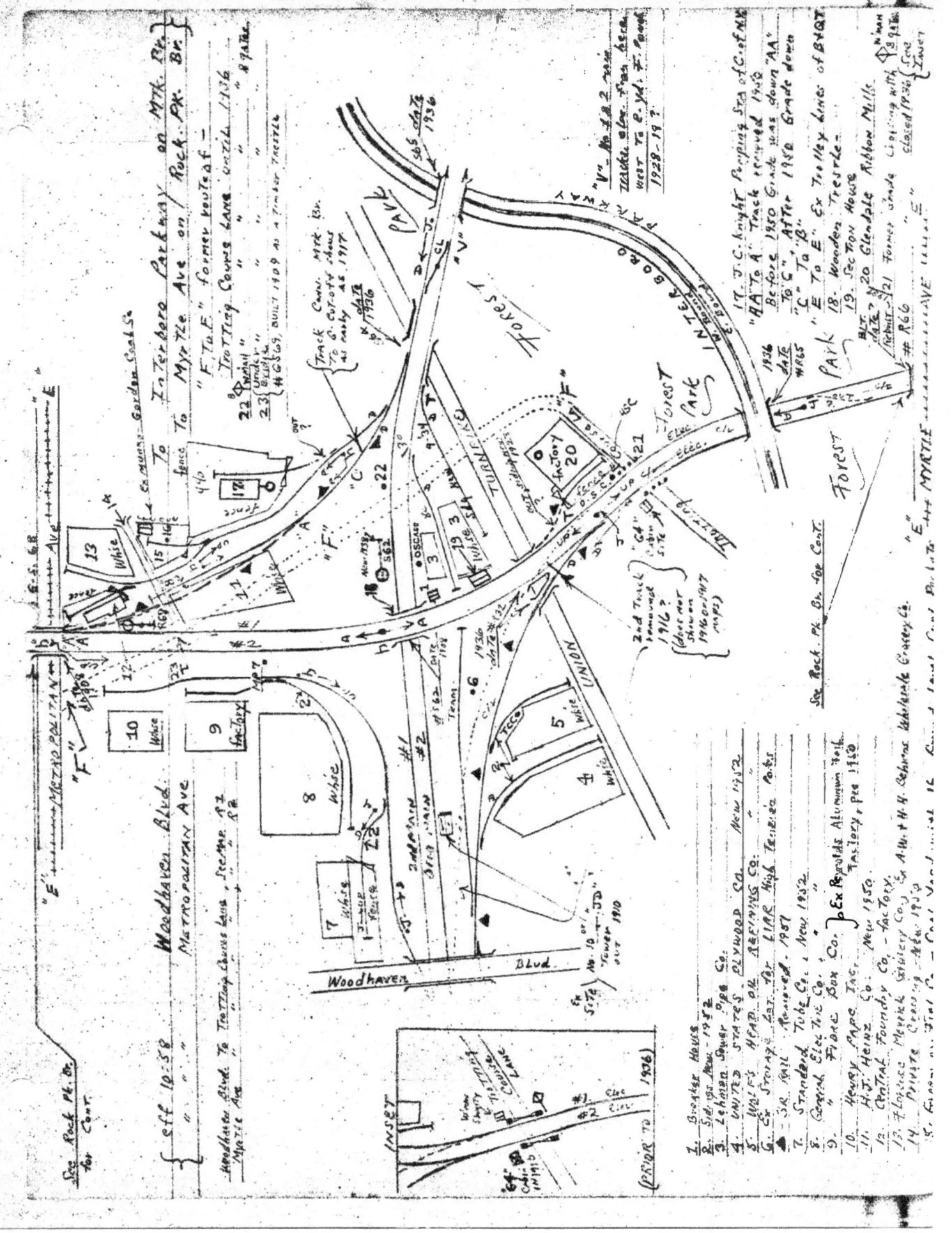

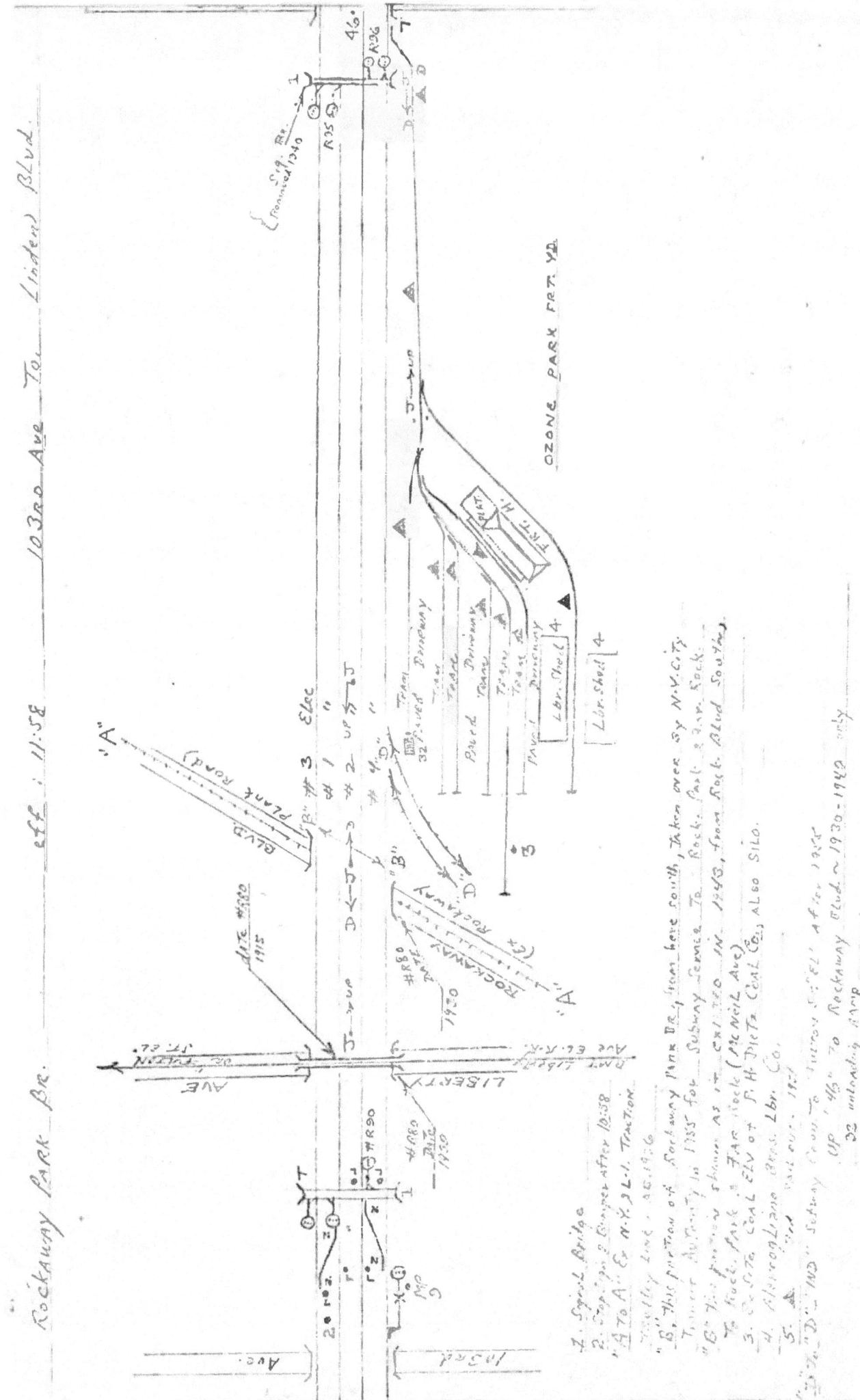

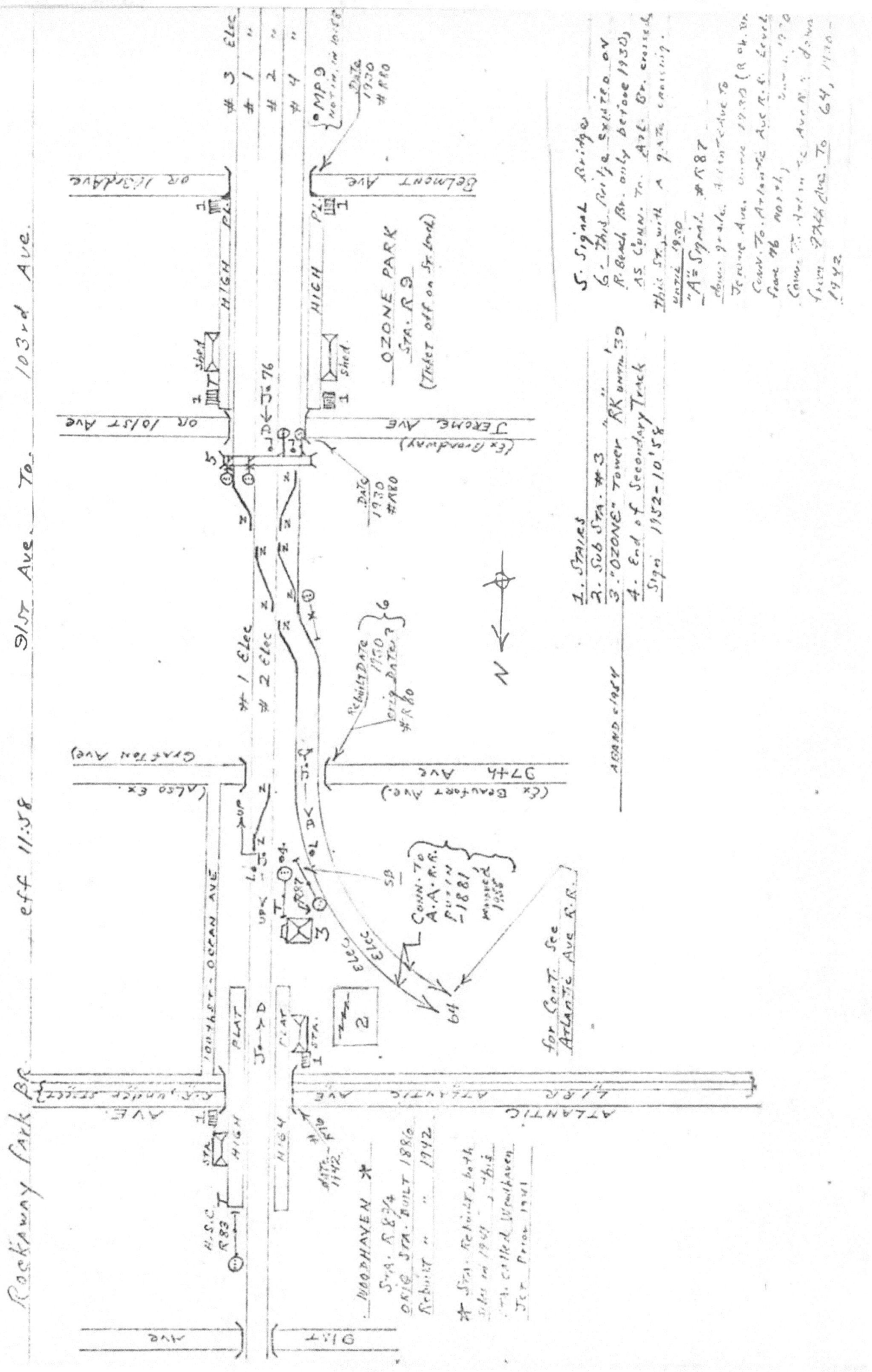

ROCKAWAY BEACH BRANCH
GLENDALE CUTOFF

Glendale cutoff built as a double-track electrified line in 1910 from White Pot (now Rego Park) to Glendale Junction.

Line opened in 1910.

New York, Woodhaven and Rockaway built double track steam road from Glendale Junction to Rockaway Park in 1880. Connection to Atlantic Branch at Woodhaven Junction built in 1881.

A single track connection built in 1887 from Hammel to NY&R.B Jct. (on Brooklyn and Montauk R.R.) by NYW&R.

New York Woodhaven and Rockaway reorganized in 1887 as the New York and Rockaway Beach Railway Co. Line was leased by LIRR in that same year. LIRR operated its trains to Far Rockaway and Rockaway Park over the NY & RB until 1901 when NY & RB was merged with LIRR.

DOUBLE TRACKING

Hammel to NY & RB Jct--1899 (by LIRR)
   3rd track added, 1904
3rd & 4th tracks, Ozone Park to Hamilton Beach--1903
BRT operated its trains over route from Woodhaven Junction to Rockaway Park from 1898 to 1917.

ELECTRIFICATION--main tracks

Woodhaven Junction to Rockaway Park--7/05
Glendale Junction to Woodhaven Junction--6/10
Hammel to NY & RB Jct.--12/05
Freight sidings--1928-30

Woodhaven Sta.
ca. 2002; Connection to
Barclay Arena at Lower Level

Ozone Pk. ca. 1962

Ozone Park Station, Circa 1970's

Ozone Park Station, Circa 1970's

Typical Section at Ozone Park

COST-EFFECTIVE PURPLE LINE SUCCEEDS IN SECURING PRIVATE-SECTOR FINANCING

Final Financial Documents Signed Today [6/17/16] on $5.6 Billion, 36-Year Contract

Maryland Department of Transportation
The Secretary's Office

Larry Hogan
Governor

Boyd K. Rutherford
Lt. Governor

Pete K. Rahn
Secretary

FOR IMMEDIATE RELEASE   Contact: Erin Henson
Teri Moss
MDOT Public Affairs
(410) 865-1025

COST-EFFECTIVE PURPLE LINE SUCCEEDS IN SECURING PRIVATE-SECTOR FINANCING

Final Financial Documents Signed Today on $5.6 Billion, 36-Year Contract

HANOVER, MD (June 17, 2016) – Maryland Transportation Secretary Pete K. Rahn announced that the Maryland Department of Transportation (MDOT), MDOT's Maryland Transit Administration (MTA), and the Purple Line Transit Partners, LLC, today signed the final financial documents on the $5.6 billion Purple Line Public-Private Partnership (P3) contract. By completing the financial close milestone, all public and private sector partners have cleared the way for the Purple Line Transit Partners to proceed to design, build, operate, finance and maintain the 16-mile light rail system from Bethesda to New Carrollton. The Purple Line Transit Partners secured all its financing for the project this week, including: an $875 million Transportation Infrastructure Finance Innovation Act loan from the U.S. Department of Transportation; $313 million in Private Activity Bonds issued by the Maryland Economic Development Corporation; and $138 million from the partners' own private equity.

"Bond investors' strong interest and our partners' private equity investments are a testament to the value and future success of the Purple Line," said Transportation Secretary Rahn. "Today's financial close keeps us on schedule with a fall construction start on the Purple Line that will connect Metro rail and bus, MARC, Amtrak and local buses into a true transportation network for Maryland residents."

This financing secures the savings MDOT anticipated by delivering a cost-effective project that reduced the original cost estimate over the life of the project from $6.2 billion to $5.6 billion. With future fare revenues and local and federal contributions, the project will cost the state less than $3.3 billion over the 36-year life of the agreement while delivering the first P3 light rail project in the nation.

"The parties that formed Purple Line Transit Partners have a strong record of constructing large projects and vast experience operating and maintaining light rail," said MTA Administrator Paul Comfort. "This contract will deliver this vital rail project that will strengthen Maryland's transportation network and will move people more efficiently."

Construction is on schedule to start late this year with service starting in spring 2022. The Purple Line will run east-west inside the Capital Beltway, with 21 stations connecting to: Metrorail's Orange, Green, and Red lines; the MARC Brunswick, Camden and Penn lines; and Amtrak at New Carrollton. Daily ridership is estimated to reach more than 74,000 by 2040. The total construction cost in the P3 contract to design and build the Purple Line is $2 billion.

"Purple Line Transit Partners is pleased to reach financial close on this important project," said Herb Morgan, CEO, Purple Line Transit Partners. "This milestone, plus the recent Board of Public Works action approving the P3 contract, solidifies Maryland's leadership in protecting and enhancing the state's fiscal integrity by advancing a public-private partnership project that will transfer construction, operation, maintenance and performance risks to the private partners all while ensuring riders and stakeholders receive improved mobility, environmental compliance and safety. Our team is looking forward to working with Maryland communities to start construction later this year and deliver this innovative project at a fixed-price and on schedule."

Purple Line Transit Partners is led by three experienced and successful P3 developers and equity investors. Meridiam Infrastructure Purple Line, LLC (Meridiam), Fluor Enterprises Inc. (Fluor) and Star America Purple Line, LLC (Star America), all Equity Members of Purple Line Transit Partners have provided the required equity investment for the Purple Line Project. The Purple Line Transit Partners headquarters are located in Riverdale, Maryland, in Prince George's County.

The federal government has reserved $900 million for the Purple Line in the New Starts Program. MTA is continuing to work with the U.S. Department of Transportation's Federal Transit Administration on a Full Funding Grant Agreement with the hope of finalizing the agreement this summer.

The Purple Line Transit Partners will manage a team of primarily local workers and contractors to ensure successful, creative and efficient delivery of the project. MTA will own the Purple Line and continue to set fares and collect fare revenue. More information on the project can be found at purplelinemd.com.

###

Home   News   Weather   Investigations   Entertainment

COVERING PRINCE GEORGE'S COUNTY

TRACEE WILKINS AND THE NEWS4 TEAM COVERING WHERE YOU LIVE

Construction Begins on Maryland's Purple Line

The line will cost about $5.6 billion and create 52,000 jobs, planners say.

By Sophia Barnes

Construction began Monday on the Purple Line, which has received mixed responses from local residents. (Published Monday, Aug. 28, 2017)

TRENDING STORIES

1 Diplomat's Daughter Stabs Boy at School in Georgetown

Construction Begins on Maryland's Purple Line - NBC4 Washington

Construction began Monday on the Purple Line, which has received mixed responses from local residents. (Published Monday, Aug. 28, 2017)

Maryland Man Reunites With Birth Mother After 40 Years

No Prison Time for Officer Who Struck Suspect With Cruiser

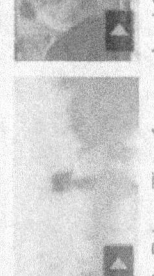
Iraqi Officials Visit Laurel, Meet With Leaders

Daughter Raises Awareness After Dad's Sudden Death

2	Florida Braces for 'Extremely Dangerous' Irma
3	Powerful Hurricane Irma Bears Down on Caribbean Islands
4	Trump Orders End to Program Protecting Immigrant 'Dreamers'
5	Immigrants Are Sought for Labor Shortage in Harvey Recovery

WEATHER FORECAST

Washington, DC

 71° Overcast
Feels Like 71°

Radar Forecast Maps

WHAT DO YOU THINK?

If you have a staycation day do you end up doing housework?

○ Yes
○ No

Construction started Monday on the long-awaited Purple Line after Maryland Gov. Larry Hogan hosted a groundbreaking ceremony for the light-rail project.

The 16 miles of track will connect Montgomery and Prince George's counties. News4's Tracee Wilkins reported that the line will serve 21 stations, connecting the Bethesda and New Carrollton Metro stops.

Planners say the Purple Line project will create 52,000 new jobs in Maryland.

The project is expected to cost $5.6 billion and will be funded through a public-private partnership. Private sector partners will contribute about $550 million, Gov. Hogan said. The federal government recently announced a $900 million contribution.

The line is not part of the Metro system.

The groundbreaking ceremony took place at the Purple Line operations center in Hyattsville, Maryland. Governor Hogan appeared with U.S. Secretary of Transportation Elaine Chao.

The purple line is "a good example of what can be accomplished when federal, state and local partners work together," Chao said.

Commuters may take up to 41,000 daily trips on the Purple Line by 2035, Chao

http://www.nbcwashington.com/news/local/Construction-Begins-on-Purple-Line-Maryland-hogan-new-metro-442009483.html[9/6/2017 12:20:57 AM]

Construction Begins on Maryland's Purple Line - NBC4 Washington

said.

Opponents have vowed to continue fighting against the Purple Line, News4's Adam Tuss reported. The project was delayed by one federal judge in May, but later that month the lawsuit was dismissed.

Published at 11:28 AM EDT on Aug 28, 2017 | Updated at 6:07 PM EDT on Aug 28, 2017

Next

Insights powered by CivicScience | Privacy Policy

4 RESPONDS
CONSUMER COMPLAINT? CLICK HERE NOW

NEWSLETTERS

Receive the latest local updates in your inbox

Email | Sign up

Privacy policy | More Newsletters

You May Like

Promoted Links by Taboola

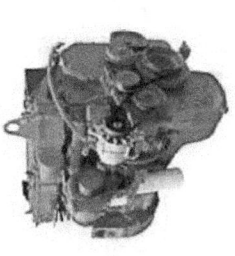
Triplets Take A DNA Test, Only To Be Surprised By Tr...
worldemand

Colin Kaepernick's Mom Expresses Outrage By Her ...
SportsChew

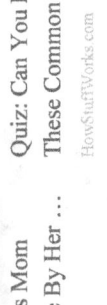
Quiz: Can You Diagnose These Common Car Proble...
HowStuffWorks.com

Opponents Fight Back Against Multi-Year Closure of Georgetown Branch Trail

'My Heart Is Devastated': Godmother of 19-Year-Old Woman Killed Mourns, Call...

Utah Nurse's Arrest Raises Questions on Evidence Collection

SPONSORED LINKS

- White-Haired Baby Surprises Doctors Until They Realize (HealthSkillet)

MORE FROM NBC Washington

- Missing Temple Student Found Dead; Fmr. Student Charged With Murder

BETHESDA | CHEVY CHASE | GAITHERSBURG | KENSINGTON | NORTH BETHESDA | POTOMAC | ROCKVILLE | SILVER SPRING

State, Purple Line Builders Sign Financing Agreements

The private team tasked with building and operating the light-rail line has lined up more than $1.3 billion for the project

BY ANDREW METCALF

Published: 2016.06.20 10:23

With money in hand, the private team tasked with building the Purple Line is expected to keep to its schedule to start construction this fall.

On Friday, Purple Line Transit Partners and the Maryland Transit Administration (MTA) announced they have signed financing agreements for the $5.6 billion, 36-year contract to construct the 16.2 light-rail line that will connect Bethesda with Prince George's County.

Purple Line rendering
MARYLAND TRANSIT ADMINISTRATION

Purple Line Transit Partners, which is comprised of the construction firm Fluor Enterprises and the financial firms Meridiam and Star America, announced it had secured about $1.3 billion in funding. Most of that amount—$875 million—came from a federal transportation loan, while another $367 million was provided through a "Green Bond" underwritten by JP Morgan and RBC Capital Markets, according to Meridiam. An additional $138 million was provided by the three firms that make up the private partnership.

In a press release, Meridiam said the bonds were sold "at the lowest rates ever achieved" in the U.S. public-private-partnership (P3) market.

"Projects like the Purple Line are extremely encouraging markers in the evolution of the U.S. P3 market," Thierry Deau, Meridiam's CEO and founder, said in the release. "Our U.S. team looks forward to the opportunity of delivering a best practice example of light-rail transit to the full spectrum of Maryland Purple Line stakeholders, beginning with the communities served by the project."

"Today's financial close keeps us on schedule with a fall construction start on the Purple Line that will connect Metro rail and bus, MARC, Amtrak and local buses into a true transportation network," Transportation Secretary Pete Rahn said.

All that is left to secure in terms of financing is a $900 million grant from the Federal Transit

Administration, which is scheduled to be transferred to the project in late July or early August after the federal grant agreement is finalized, according to MTA.

That grant will give the construction team about $2.2 billion to work with as it constructs the rail line over the next five years. The project is estimated to cost about $2 billion to construct, according to the MTA.

The loans will be repaid by Purple Line Transit Partners using the state's monthly $150 million installments, which will be provided over the life of the 36-year contract. Those monthly payments will also be used by Purple Line Transit Partners to operate and maintain the light-rail line until it's turned over to the Maryland Transit Administration at the end of the contract.

What's not yet known is whether a federal lawsuit brought by Chevy Chase residents will impact the construction timeline of the project and in turn the loans provided to build it. On Thursday a federal judge considered possibly delaying the project to allow an investigation into whether well-documented recent problems with the Metro system could negatively impact ridership of the Purple Line. A lawyer for the state told U.S. District Court Judge Richard Leon a delay could allow investors to pull out from the project. Leon gave both sides two more weeks to provide additional statements about the case.

Back to Bethesda Beat

7 Comments   Bethesda Magazine   Login

Recommend   Share   Sort by Best

Join the discussion…

Bethesda Mom · a year ago
Bethesda Beat you have a typo it's Fluor not Flour (Fluor Enterprises Inc.)
Reply · Share ›

Bethesda Beat > Bethesda Mom · a year ago

Attachments

- image003.jpg (22.71KB)
- image004.png (3.31KB)

 Maryland Department of Transportation

Fast Facts on the Purple Line Project
March 2016

Purple Line Basics

- The Purple Line is a 16.2-mile light rail line between Bethesda in Montgomery County and New Carrollton in Prince George's county.
- The Purple Line will have 21 stations with stops in Silver Spring, Takoma/Langley Park and the University of Maryland at College Park.
- It will connect with 4 branches of the Metrorail system at Bethesda, Silver Spring, College Park, and New Carrollton; as well as all three MARC commuter rail lines, and Amtrak.
- Modern light rail is a safe, quiet, electrically-powered transit vehicle that can operate in existing streets or in its own right-of-way.
- At opening, trains will operate every 7 ½ minutes during peak periods, 10-15 minutes off peak.
- Daily ridership on the Purple Line is projected to reach 74,000 by 2040.
- Construction begins in late 2016.
- Service will begin in Spring 2022.

Economic Benefits

- Over 6,300 jobs will be created over the course of Purple Line construction
- Job center in Prince George's County
- Bonding program support for DBEs
- Purple Line DBE goals of 22% for construction and 26% for design are more than double the federal requirements of 10%
- MTA will connect the Concessionaire with local DBEs for contracting opportunities
- Rail transit investments are typically catalysts for revitalization and development
- Improved access to the regional rail system means access to more jobs for local residents

Operations and Maintenance Facility in Prince George's County

Environmental Benefits

- The Purple Line will take 17,000 cars off the road every day, saving 1 million gallons of gas annually
- Electric power means no air emissions into the immediate environment
- Purple Line's use of existing roadways minimizes effects on land and water resources

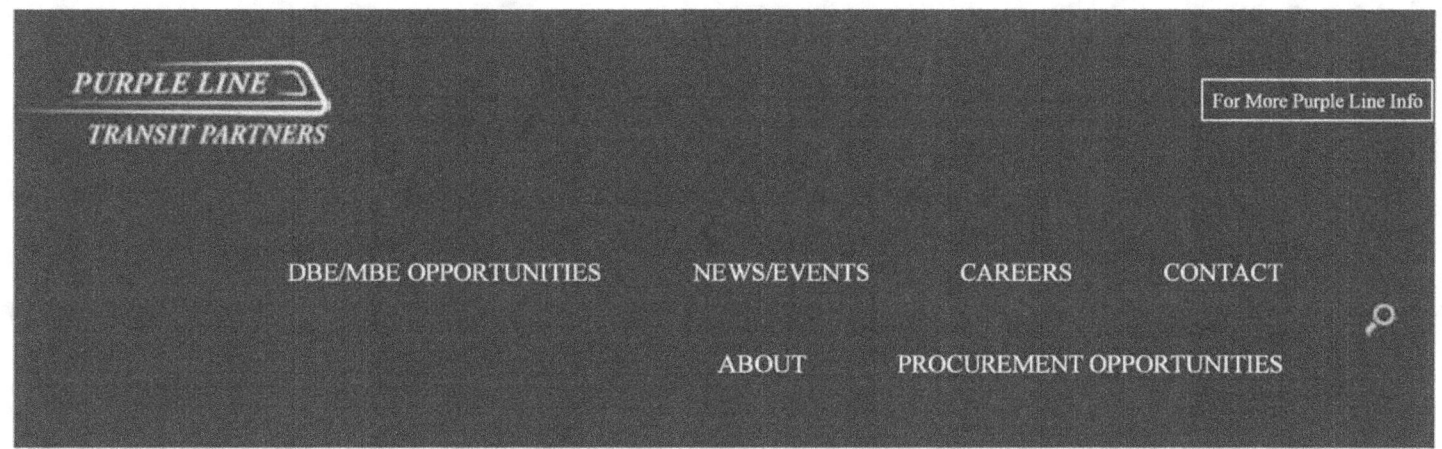

Meet The Team: Purple Line Transit Partners

ABOUT

PROJECT OVERVIEW

MEET THE TEAM

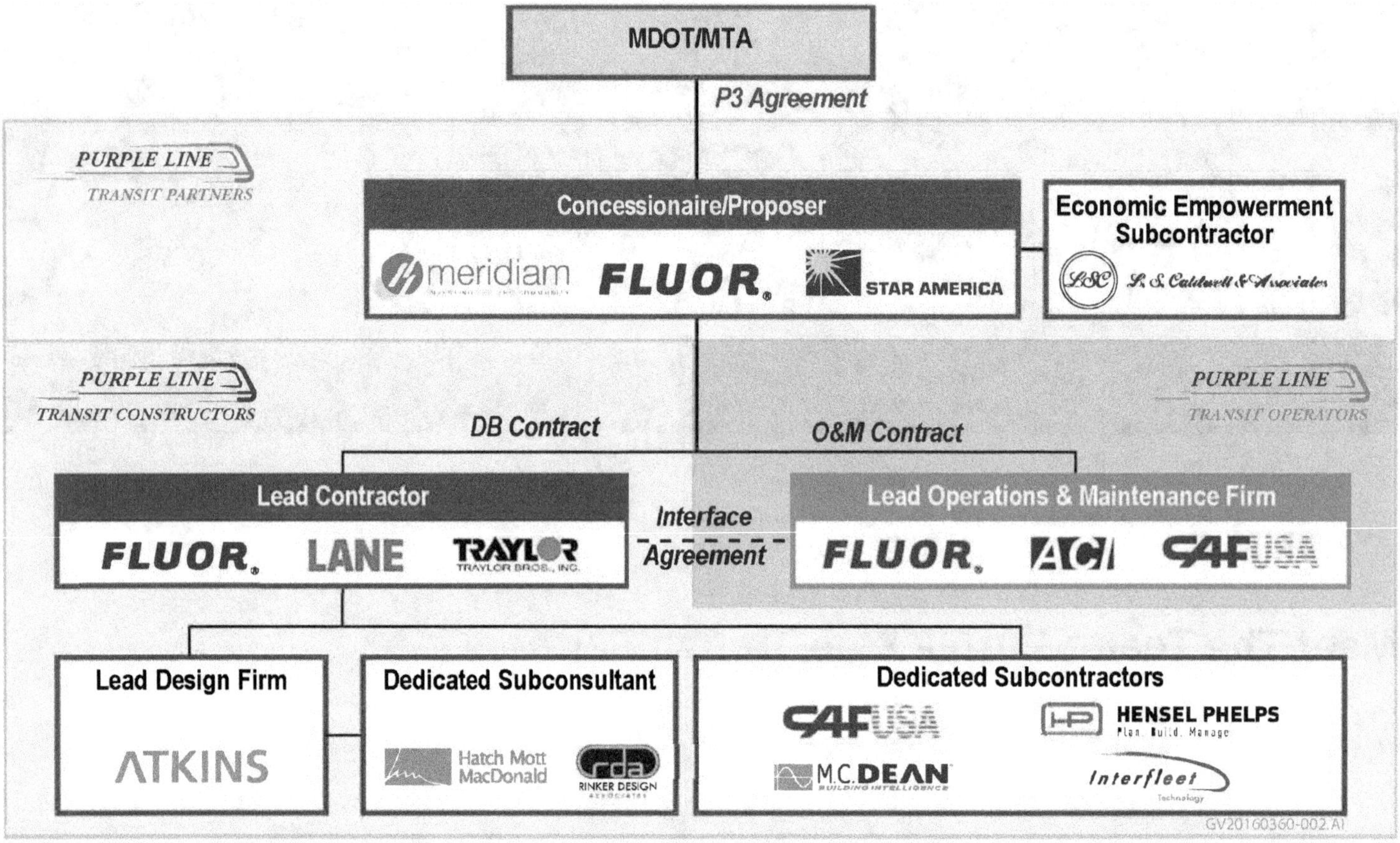

Meet The Team | Purple Line Transit Partners

[Organizational chart showing: MDOT/MTA at top connected via P3 Agreement to Concessionaire/Proposer (Meridiam, FLUOR, STAR AMERICA) under Purple Line Transit Partners, with Economic Empowerment Subcontractor (S.S. Caldwell & Associates). Below, DB Contract connects to Lead Contractor (FLUOR, LANE, TRAYLOR BROS., INC.) under Purple Line Transit Constructors, and O&M Contract connects to Lead Operations & Maintenance Firm (FLUOR, ACI, CAF USA) under Purple Line Transit Operators, linked by Interface Agreement. Lead Contractor connects to Lead Design Firm (ATKINS), Dedicated Subconsultant (Hatch Mott MacDonald, Rinker Design Associates), and Dedicated Subcontractors (CAF USA, HENSEL PHELPS, M.C. DEAN, Interfleet Technology).]

Purple Line Transit Partners (PLTP) is led by three experienced and successful public-private partnership (P3) developers and equity investors that have ample financial capacity to fully commit the equity required for a project of this nature and size and to fund such equity when required. Meridiam Infrastructure Purple Line, LLC (Meridiam), Fluor Enterprises Inc. (Fluor) and Star America Purple Line, LLC (Star America), all Equity Members of PLTP, will provide the required equity investment for the Purple Line Project.

The combination of PLTP's Equity Members represents an investment approach that provides for alignment of interests within the consortium through vertical integration and a long-term view from inception all the way through to hand-back. Fluor's involvement as an Equity Member, a member of the Design-Build Joint Venture, and a member of the O&M Joint Venture, has ensured that one of its members is involved in every phase of the Project, guaranteeing that the interests of the entire consortium will be aligned.

TEAM MEMBERS

Meridiam is a leading equity investor, developer, asset manager, and long-term partner in P3 projects in the U.S., Canada, and Europe. Meridiam's dedicated 25-year funds enable it to be a long-term partner with the public sector from project inception through operations. With global assets under management, including rail projects, of approximately $5 billion (representing a construction value in excess of $40 billion), Meridiam has a distinct position in the industry as a community-values investor in P3 transportation projects that involve new construction or capital improvements. Meridiam's investors include public pension funds, labor pension funds, and life insurance companies.

Fluor is the leading design-build investor in the U.S. market, having developed the first P3 transportation projects in Virginia, South Carolina, Colorado, and Texas. Fluor is the largest publicly traded engineering / procurement / construction company in the U.S. and has the highest long-term debt ratings in the industry. Fluor approaches project equity investment as part of a whole life approach to projects which includes development, financing, design, construction, O&M, and rehabilitation for transportation P3 projects. Fluor has current cash and investments in excess of $3 billion and committed credit facilities of $2.8 billion to source equity investment funding requirements. Fluor brings vast urban transit and world-class P3 experience to this project.

Star America is a U.S.-based independent developer, investor, and manager of public infrastructure, focusing primarily on greenfield P3 projects in North America. Star America team members have invested in or managed more than 33 infrastructure projects valued at more than $28 billion, as well as advised on more than 25 P3 transactions, a majority of those projects in the transportation sector.

Purple Line Transit Constructors (PLTC)

The design-build team of Fluor, The Lane Construction Corporation (Lane), and Traylor Bros., Inc. (Traylor) combines companies with experience as design-build delivery contractors on dozens of applicable, mega design-build projects. Fluor and Lane, in a number of different capacities, have worked on many major design-build projects within the region and throughout the United States. Traylor is an industry-leading civil and tunnel construction company with proven expertise in urban transit environments. The team also includes the local knowledge, experience, and execution capabilities of PLTC's dedicated lead design firm Atkins and dedicated design sub-consultant Hatch Mott McDonald, and construction subcontractors M.C. Dean and Hensel Phelps, as well as a group of dedicated subcontractors, including Light Rail Vehicle (LRV) supplier CAF and Interfleet, a Rolling Stock Advisor.

Purple Line Transit Operators (PLTO)

PLTO will deliver the O&M services for the Purple Line System. The PLTO member entities of Fluor, Alternate Concepts Inc. (ACI), and CAF USA, Inc. (CAF) possess extensive experience to draw upon for the Purple Line. Fluor maintains rail systems in the Netherlands that are part of the Trans-European rail network. Fluor and ACI are teamed to provide 30 years of O&M services on the Denver Eagle P3 Project. ACI has 25 years of experience operating and maintaining transit services throughout North America, including direct operation of rail services in Phoenix, Boston, and Puerto Rico. CAF maintains more than 4,000 vehicles located in Houston, Texas; Zaragoza, Spain; Mexico City, Mexico; and Sydney, Australia; and operates a full-service assembly and manufacturing plant in Elmira, New York.

PLTO will integrate the knowledge base of these firms with the expertise gained from rail start-ups, such as the MBTA Commuter Rail (2003), Tren Urbano Heavy Rail (2004), Valley Metro Light Rail (2008), Houston Metro (2013), and others to achieve the long-term success of the Purple Line.

PLTO is engaged from the development phase of the project through substantial completion. Resources are available at each phase of the project beginning with inclusion in the Technical Working Groups where PLTO representatives provide input into operational design elements, constructability and operability reviews, and service life analysis of the components of the system. During the construction and pre-revenue phases, PLTO will provide design reviews for maintainability and operability, support for plan and procedure preparation, and the training for vehicle and system maintenance. Operators will also be hired and trained from the local labor force for the vehicles to be burned in pre-revenue. At substantial completion, PLTO will assume operational responsibility in accordance with the project agreement.

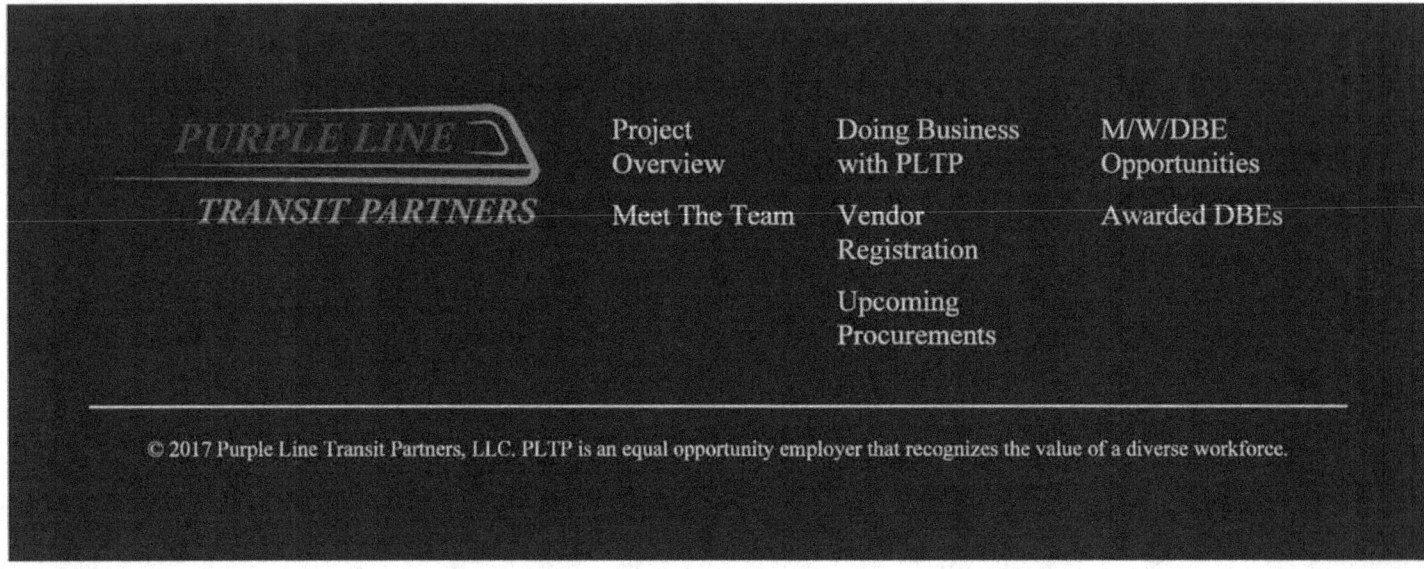

Project Overview

Overview

Fast Facts for the Purple Line Project

The Purple Line is a 16-mile light rail line that will extend from Bethesda in Montgomery County to New Carrollton in Prince George's County. It will provide a direct connection to the Metrorail Red, Green and Orange Lines; at Bethesda, Silver Spring, College Park, and New Carrollton. The Purple Line will also connect to MARC, Amtrak, and local bus services. Twenty-one stations are planned.

The Purple Line will be light rail and will operate mainly in dedicated or exclusive lanes, allowing for fast, reliable transit operations.

MTA is taking the lead on this project, with the support and close coordination of a team that includes the Washington Metropolitan Area Transit Authority, Montgomery and Prince George's counties, the Maryland-National Capital Park and Planning Commission, State Highway Administration, and local municipalities in the project area.

Project Overview - Maryland Purple Line

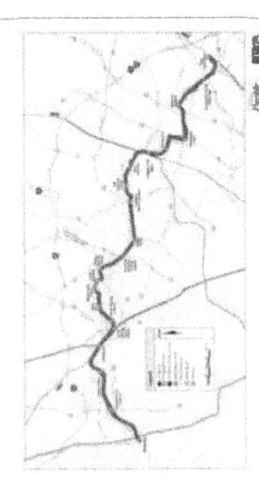

The Purple Line is a 16-mile east-west light rail line linking Bethesda, Silver Spring, Takoma/Langley Park, the University of Maryland at College Park, and New Carrollton.

Benefits

- Reliable and rapid east-west travel
- Connects to Metrorail Green and Orange lines and both branches of the Red Line
- Supports community revitalization and transit-oriented development
- Connects people to jobs
- Serves major economic centers
- Connects to all three MARC lines, Amtrak, and local bus routes

Highlights

- 16.2 miles
- 1 short tunnel (Wayne Avenue to Long Branch)
- 21 stations
- 69,000 total daily riders in 2030; 74,000 total daily riders in 2040
- Construction: 2016-2021

The project includes the completion of the Capital Crescent Trail between Bethesda and Silver Spring, the completion of the Green Trail along Wayne Avenue to Sligo Creek Parkway, and the construction of a bike path through the University of Maryland Campus.

Milestones

2002-2008	MTA studies a range of alignments and different transit modes (light rail and bus rapid transit) for the Purple Line project area
2009	Light rail selected as the mode of transit; alignment identified
2009-2014	Conceptual and Preliminary Engineering Phase
2013	MTA decides to use a Public-Private Partnership (P3) to design, build, finance, operate and maintain the Purple Line
2013	FTA accepts the Final Environmental Impact Statement
2014	FTA issues the Record of Decision
2014	MTA issues Request for Proposals for Purple Line P3
2016	Concessionaire selected to complete design, build, operate and maintain the Purple Line

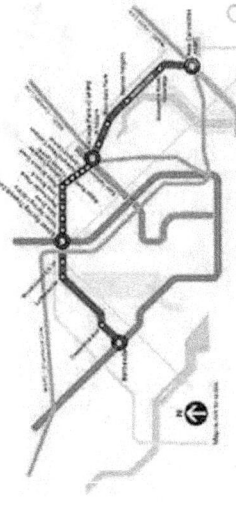

Community-Friendly Light Rail Transit

Light rail vehicles are modern streetcars, powered by overhead electrical wires.

Features include:

- Low floors to allow passengers to board without climbing steps
- Quiet operations
- Neighborhood stations convenient for pedestrians

2016-2021   Final Design and Construction

2022   Purple Line service begins

Purple Line Information
443-451-3706
443-451-3705 (Español)

Search   Terms of Use   MTA Notice of Compliance   Title VI   Site Map   Login

Maryland Transit Administration
5811 Kenilworth Avenue, Suite 3A
Riverdale, Maryland 20737

Public-Private Partnership

Maryland DOT and the Maryland Transit Administration pursued an innovative solicitation approach with the Purple Line. The State has solicited a single private partner (concessionaire) who will be responsible for designing, constructing, operating, and maintaining the project, and the private partner will also help finance a portion of construction. On March 2, 2016, Governor Larry Hogan announced Purple Line Transit Partners as the concessionaire for the Purple Line. This partnership with private industry, commonly referred to as a public-private partnership (PPP or P3), will promote the successful delivery of the Purple Line. P3 Contract provided below.

Web server limitations prevent us from listing all of the documents which are referenced or referred to in the documents on this page. If you wish to review a reference document referred to in the documents, you may request that it be provided to you. Please do so by contacting outreach@purplelinemd.com and including the name(s) of the specific document(s) you are requesting in your email.

Execution Version – March 2, 2016

- Public-Private Partnership (P3) Agreement with Exhibits
- Technical Provisions Book 2 – Part 1
- Technical Provisions Book 2 – Part 2
- Technical Provisions Book 2 – Part 3
- Codes and Standards Book 3
- Contract Drawings Book 4 – Plans *(Warning: Large file)*
- Contract Drawings Book 4 – Additional Plans
- Contract Drawings Book 4 – Right of Way Plans *(Warning: Large file)*

While Book 4 consists of the Contract Drawings referenced by the P3 Agreement, the Concessionaire will continue to work alongside MTA/MDOT after award to further develop

Public-Private Partnership - Maryland Purple Line

and complete the design as the project progresses.

Report to the Maryland General Assembly: Public-Private Partnership for the Purple Line – Description of the Proposed P3 Agreement

- Appendix 1 – Risk Allocation
- Appendix 2 – P3 Agreement Policy Guide

Fast Facts on the Purple Line Public-Private Partnership

Fast Facts on Third Party Projects

Purple Line P3 FAQs

News Release – P3 Announcement – March 2, 2016

FaLang translation system by Faboba

Purple Line Information
443-451-3706
443-451-3705 (Español)

Search   Terms of Use   MTA Notice of Compliance   Title VI   Site Map   Login

Maryland Transit Administration
6811 Kenilworth Avenue, Suite 3A
Riverdale, Maryland 20737

Kansas City Streetcar Funding Method
"Transit Improvement District"

Create a "Transit Improvement District"

Example: Kansas City, MO.

The city advances the funding for the construction of the new streetcar line in the form of Bonds, then recoups the funds:

Recently, by nearly a 2/3 majority, voters within the "Transit Improvement District" approved a 1% sales tax, and a modest real estate tax increase to fund construction and operation of their new streetcar project. The project is also funded by parking assessments and federal funding.

Construction for the Downtown Kansas City streetcar starter line officially begins in Spring 2014. The completed starter line will include a two mile round trip streetcar route (four miles of track) along Main Street connecting Kansas City's River Market area to Crown Center and Union Station. It will serve the city's Central Business District, the Crossroads Art District, the Power and Light District and numerous other businesses, restaurants, art galleries, educational facilities and residential neighborhoods. The starter line will include 16 stops spaced approximately every two blocks. Additionally, the Singleton Yard Streetcar Vehicle Maintenance Facility and Park & Ride lot will be built at 3rd Street and Grand Blvd. in the River Market.

Construction will take approximately 18 months and create hundreds of local jobs. The KC Streetcar Constructors and the City of Kansas City are committed to minimizing impacts to those who live and work downtown during construction.

The Downtown KC Streetcar starter line is the next step in a longer-range plan to create a regional, integrated transit system to uniquely connect the Greater Kansas City area like never before. Progressive regions around the country with streetcar systems have seen significant economic growth and the Downtown KC Streetcar starter line is a step in effort to realize an even more vibrant, vital and livable urban center. Streetcar systems attract new residents, businesses and workforce and provide an improved and more efficient travel option. It is envisioned that the downtown KC Streetcar starter line will bring new investment and increased property values to downtown along with an increased economic impact during construction and after.

The completion of the Downtown KC Streetcar starter line project is anticipated in summer of 2015 followed by a period of testing. It is expected that by the end of

Funding by Assessment on Property Owners

P. S. C. ORDERS WORK ON UTICA AVE. LINE

Brooklyn Daily Eagle May 18, 1915

First Subway to Be Built by Assessment on Property Owners.

The Public Service Commission today ordered its engineering staff to commence work at once on making plans for the Utica avenue subway line from Eastern Parkway to Flatbush avenue. This action was taken upon a motion by Commissioner Williams, the vote being unanimous.

This is the first official step by the Commission on following the passage by the last Legislature of the enabling act under which this line may be contructed by assessment upon the affected property owners. The formal application from the property owners that the city proceed to build on the assessment plan was received some months ago.

The plans for the Eastern Parkway line call for a turnout at Utica avenue to connect with this line. It was stated at the Commission's office today that work on the plans would be commenced at once and pushed along with all possible speed. This will be the first line to be constructed by assessment in the city's history.

LAWS

OF THE

STATE OF NEW YORK,

PASSED AT THE

ONE HUNDRED AND THIRTY-EIGHTH SESSION

OF THE

LEGISLATURE,

BEGUN JANUARY SIXTH, 1915, AND ENDED APRIL TWENTY-FOURTH, 1915,

AT THE CITY OF ALBANY,

AND ALSO OTHER MATTERS REQUIRED BY LAW TO BE PUBLISHED WITH THE SESSION LAWS.

Vol. II.

ALBANY
J. B. LYON COMPANY, STATE PRINTERS
1915

the case of any such sale or conveyance accept in part payment for the property sold a bond or other obligation to the city secured by purchase money mortgage on said property, such bond or other obligation and such mortgage to contain such terms and conditions as the commission may deem proper, including in the discretion of the commission provision for the payment of the amount of said bond or other obligation in installments.

§ 3. This act shall take effect immediately.

Chap. 545.

AN ACT to amend section thirty-seven of chapter four of the laws of eighteen hundred and ninety-one, entitled "An act to provide for rapid transit railways in cities of over one million inhabitants," with reference to assessment of cost and expense necessary to be incurred for the construction of a rapid transit railroad and for property to be acquired for the construction and operation thereof upon property benefited thereby.[1]

Became a law May 8, 1915, with the approval of the Governor. Passed, three-fifths being present.

Accepted by the City.

The People of the State of New York, represented in Senate and Assembly, do enact as follows:

L. 1891, ch. 4, § 37, subd. 3, as added by L. 1894, ch. 752, and amended by L. 1895, ch. 519, L. 1904, ch. 562, L. 1906, chaps. 472, 607, L. 1907, ch. 534, L. 1908, ch. 472, L. 1909, ch. 498, and L. 1911, ch. 888, amended.

Section 1. Subdivision three of section thirty-seven of chapter four of the laws of eighteen hundred and ninety-one, entitled "An act to provide for rapid transit railways in cities of over one million inhabitants," as added by chapter seven hundred and fifty-two of the laws of eighteen hundred and ninety-four, and amended by chapter five hundred and nineteen of the laws of eighteen hundred and ninety-five, chapter five hundred and sixty-two of the laws of nineteen hundred and four, chapter four hundred and seventy-two of the laws of nineteen hundred and six, chapter six hundred and seven of the laws of nineteen hundred and six, chapter five hundred and thirty-four of the laws of nineteen hundred and seven, chapter four hundred and seventy-two of the laws of nineteen hundred and eight, chapter four hundred and ninety-eight of the laws of nineteen hundred and nine, chapter eight hundred

---

[1] The amendments effected by this act are so numerous and extensive that it is impracticable to indicate the changes made.

and eighty-eight of the laws of nineteen hundred and eleven and chapter five hundred and forty of the laws of nineteen hundred and thirteen,[2] is hereby amended to read as follows:

3. The words "rapid transit railroad," "railroad," "road," or "improvement" as hereinafter in this section used shall severally include a rapid transit railroad, and any part thereof, and any improvement or addition thereto, that shall be the subject of action hereunder, and shall severally include any and all property, including equipment other than rolling stock, that shall be necessary either for the construction or the operation of such a rapid transit railroad. A rapid transit railroad owned or to be owned by the city, and for the construction of which with public money in whole or in part a contract or contracts have been or are authorized by this act to be entered into as aforesaid, shall be a local improvement the cost of which railroad may be met in whole or in part by assessment on the property benefited. The public service commission with the approval of the board of estimate and apportionment or other analogous local authority of the city in which such rapid transit railroad is authorized to be constructed shall have power to determine whether all or any, and if any, what portion of the cost and expense necessary to be incurred for any such road shall be assessed upon property benefited thereby, and whether all or any, and if any, what portion of the cost and expense necessary to be incurred, or which shall have been already necessarily incurred, for the acquisition of any property for the construction or operation of said railroad shall be assessed upon property benefited by said railroad. An assessment or assessments upon the property so benefited may be laid, confirmed, enforced and collected in accordance with such determination and pursuant to the provisions of the charter and laws respecting assessments for local improvements in such city.

§ 2. Subdivision four of section thirty-seven of chapter four of the laws of eighteen hundred and ninety-one, entitled "An act to provide for rapid transit railways in cities of over one million inhabitants," as added by chapter seven hundred and fifty-two of the laws of eighteen hundred and ninety-four, and amended by chapter five hundred and nineteen of the laws of eighteen hundred and ninety-five, chapter five hundred and sixty-two of the laws of nineteen hundred and four, chapter four hundred and seventy-two of the laws of nineteen hundred and six, chapter six hundred and

---

[2] Subd. 3 was not affected by L. 1913, ch. 540.

L. 1908, ch. 472, L. 1909, ch. 498, and L. 1911, ch. 888, amended.

seven of the laws of nineteen hundred and six, chapter five hundred and thirty-four of the laws of nineteen hundred and seven, chapter four hundred and seventy-two of the laws of nineteen hundred and eight, chapter four hundred and ninety-eight of the laws of nineteen hundred and nine, chapter eight hundred and eighty-eight of the laws of nineteen hundred and eleven and chapter five hundred and forty of the laws of nineteen hundred and thirteen,[3] is hereby amended to read as follows:

Estimate of cost of construction and acquisition of property.

4. At any time after the consents have been obtained for any such rapid transit railroad and the detailed plans and specifications therefor have been prepared as hereinbefore authorized and directed, the public service commission may certify and transmit to said board of estimate and apportionment or such other analogous local authority of such city an estimate of the cost and expense necessary to be incurred for the construction of said railroad, and for the acquisition of any property, including equipment other than rolling stock, that shall be necessary either for the construction or the operation of such railroad, or from time to time an estimate of the cost and expense necessary to be incurred or a statement of the cost and expense which has been necessarily incurred for the acquisition of any property for the construction or operation of said railroad. With such estimate or statement the commission shall transmit a statement which shall show (1) the proportion of said cost and expense, together with the amount thereof in money, which should be assessed upon the property benefited; (2) the boundaries of the district or districts in said city upon which an assessment or assessments aggregating said amount should in the opinion of the commission be levied, and (3) the amount so to be levied in every such district. Thereupon the public service commission with the approval of the board of estimate and apportionment or other such analogous local authority of said city shall have power to, and, if in their judgment the interests of the public so require they shall after publishing a notice at least one week in advance in the City Record and in such other newspapers published in said city as said board of estimate and apportionment, or other local authority, shall designate as sufficient, stating the time, place and subjects to be considered, and after a joint hearing, pursuant to such notice by and before said commission and said board or other authority, which may be adjourned from time to time, in accordance with the charter and

Statement as to assessments.

Determination of property benefited and proportion of assessments.

---

[3] Subd. 4 was not affected by L. 1913, ch. 540.

laws aforesaid, fix and determine the boundaries of the district or districts upon which said assessment or assessments shall be levied, the whole amount or proportion of any such cost and expense to be assessed upon property benefited by said improvement, and the amount or proportion of such whole assessment to be levied in said district or districts respectively benefited by said improvement, and take such other and further proceedings as shall be necessary to levy and collect such assessment or assessments. The decision of said public service commission aforesaid, so approved by the board of estimate and apportionment or other such analogous local authority, shall be final as to each matter so fixed and determined and shall not be subject to review. *Decision final.*

§ 3. Subdivision seven of section thirty-seven of chapter four of the laws of eighteen hundred and ninety-one, entitled "An act to provide for rapid transit railways in cities of over one million inhabitants," as added by chapter seven hundred and fifty-two of the laws of eighteen hundred and ninety-four, and amended by chapter five hundred and nineteen of the laws of eighteen hundred and ninety-five, chapter five hundred and sixty-two of the laws of nineteen hundred and four, chapter four hundred and seventy-two of the laws of nineteen hundred and six, chapter six hundred and seven of the laws of nineteen hundred and six, chapter five hundred and thirty-four of the laws of nineteen hundred and seven, chapter four hundred and seventy-two of the laws of nineteen hundred and eight, chapter four hundred and ninety-eight of the laws of nineteen hundred and nine, chapter eight hundred and eighty-eight of the laws of nineteen hundred and eleven and chapter five hundred and forty of the laws of nineteen hundred and thirteen,[4] is hereby amended to read as follows: *L. 1891, ch. 4, § 37, subd. 7, as added by L. 1894, ch. 752, and amended by L. 1895, ch. 519, L. 1904, ch. 562, L. 1906, chaps. 472, 607, L. 1907, ch. 554, L. 1908 ch. 472, L. 1909, ch. 498, and L. 1911, ch. 888, amended.*

7. In order to provide funds in advance of the collection of such assessments, the comptroller or other chief financial officer of such city shall in addition to power to issue assessment bonds under the provisions of any law or charter of such city have also additional authority in lieu of issuing any such assessment bonds under said law or charter to issue and sell at not less than par on or after the date when any such assessment shall be confirmed and entered bonds which shall be known as rapid transit construction bonds for the railroad designated as aforesaid and which shall not exceed in the aggregate the amount of the assessment so levied as aforesaid. Except that the city may guarantee in such *Issue and sale of rapid transit construction bonds.*

City's faith and

---

[4] Subd. 7 was not affected by L. 1913, ch. 540.

bonds the validity of the assessment and the regularity of the proceedings to levy it, such rapid transit construction bonds shall not be issued or sold upon the faith or credit of the city and the faith or credit of the city shall not be pledged nor shall the city or any of the city's property be liable for the payment thereof, but such bonds shall be payable only out of the rapid transit construction fund as hereinafter directed to be constituted. Such bonds shall be in such form, denomination or denominations, and for such term, not exceeding fifteen years, as the said comptroller or other financial officer shall designate and shall bear the same rate of interest as the assessment installments shall bear. They shall be exempt from all taxation, except for state purposes, shall be receivable in payment of any such assessments or installments thereof, and may be made redeemable, in whole or in part, on any interest day after one year. They shall be a legal investment for the sinking funds of such city and for trustees and other fiduciaries charged with the investment of trust funds. If such bonds are redeemed in part, the bonds selected for redemption shall be chosen by lot, and their numbers shall be published in at least two newspapers of general circulation in such city at least twice a week for four weeks prior to the day of their redemption, and after the day specified for their redemption, the principal sums represented thereby shall bear no interest.

In selling such rapid transit construction bonds the comptroller may by the terms of sale or otherwise prescribe that payment to him therefor shall be made by the purchaser in such installments as the need of construction as certified to him by the public service commission shall require, and may provide for the forfeiture of the right to bonds allotted and of payments made thereon. All moneys derived from the sale of such bonds, and all moneys derived from the collection of such assessments shall be kept separate and apart from all other funds of the said city and shall be known as the rapid transit construction fund of such railroad. Unless the assessment be made separately for the cost and expense of acquisition of property as aforesaid, they shall be applied only to the following uses and, among such uses, only in the following order as nearly as may be: (1) To the cost and expenses of the construction of such railroad and the acquisition of property necessary for such construction, including equipment other than rolling stock; (2) to the acquisition of real property necessary for the operation thereof; (3) to the retirement of the rapid transit construction bonds therefor. In case an assessment is made separately for the cost and expense incurred or to be incurred for the acquisition of any property for the construction or operation of any such railroad, the money derived from the sale of such bonds, and all moneys derived from the collection of such assessment shall be applied only to pay or reimburse the cost and expense of acquisition of the property for which such assessment was made or to the retirement of the bonds issued in advance of the collection of such assessment.

§ 4. Subdivision eight of section thirty-seven of chapter four of the laws of eighteen hundred and ninety-one, entitled "An act to provide for rapid transit railways in cities of over one million inhabitants," as added by chapter seven hundred and fifty-two of the laws of eighteen hundred and ninety-four, and amended by chapter five hundred and nineteen of the laws of eighteen hundred and ninety-five, chapter five hundred and sixty-two of the laws of nineteen hundred and four, chapter four hundred and seventy-two of the laws of nineteen hundred and six, chapter six hundred and seven of the laws of nineteen hundred and six, chapter five hundred and thirty-four of the laws of nineteen hundred and seven, chapter four hundred and seventy-two of the laws of nineteen hundred and eight, chapter four hundred and ninety-eight of the laws of nineteen hundred and nine, chapter eight hundred and eighty-eight of the laws of nineteen hundred and eleven and chapter five hundred and forty of the laws of nineteen hundred and thirteen,[5] is hereby amended to read as follows: *L. 1891, ch. 4, § 37, subd. 8, as added by L. 1894, ch. 752, and amended by L. 1895, ch. 519, L. 1904, ch. 562, L. 1906, chaps. 472, 607, L. 1907, ch. 534, L. 1908, ch. 472, L. 1909, ch. 498, and L. 1911, ch. 888, amended.*

8. In case of default in the payment of any installment of interest or principal of any rapid transit construction bond the holder thereof may require, if necessary, by peremptory writ of mandamus, any tax lien of such city for the amount of any assessment upon the property benefited which is then due and payable, to be immediately sold or enforced in accordance with the charter and laws of such city. If at such time the tax lien so sold shall include, in addition to the lien of the assessment aforesaid, any lien for delinquent taxes or other lienable charges due to the city, and if it shall become necessary to reduce the amount of the tax lien pursuant to the charter and laws of such city, the lien shall not be reduced so as to make it less in value than the amount of the assessment aforesaid with the interest thereon, and notwithstanding any reduction as aforesaid, the proceeds of the sale of such a lien, to the extent of the full amount of the assessment and *Enforcement of tax liens in default of payments on bonds. Reduction of tax lien. Use of proceeds of lien sale.*

---

[5] Subd. 8 was not affected by L. 1913, ch. 540.

interest, shall be paid into the rapid transit construction fund of the rapid transit railroad aforesaid, and the balance, if any, shall be applied as proceeds of the rest of the tax lien.

Effect of reassessment for reduction on payments into rapid transit construction fund.

If any assessment shall be reduced for fraud, substantial error or other reason, the cost and expense aforesaid may be reassessed, and the reassessment shall stand as security for the rapid transit construction bonds aforesaid to the same degree and in the same manner as if it had been an original assessment. In case any assessment is reduced below its original amount, however, either the amount to be expended in constructing the rapid transit improvement aforesaid and for acquisition of property necessary for construction and operation thereof as aforesaid, or to be expended for acquisition of property, if the assessment reduced is for cost and expense thereof separately, shall be correspondingly reduced or else the difference between the original assessment and the reassessment shall be paid by the city into the rapid transit construction fund of the rapid transit railroad aforesaid, either from current revenue or from the proceeds of the sale of revenue bonds, corporate stock, or other obligations of such city as the board of estimate and apportionment shall determine.

L. 1891, ch. 4, § 37, subd. 9, as added by L. 1894, ch. 752, and amended by L. 1895, ch. 519, L. 1904, ch. 562 L. 1906, chaps. 472, 607, L. 1907, ch. 534, L. 1908, ch. 472, L. 1909, ch. 498, and L. 1911, ch. 888, amended.

§ 5. Subdivision nine of section thirty-seven of chapter four of the laws of eighteen hundred and ninety-one, entitled "An act to provide for rapid transit railways in cities of over one million inhabitants," as added by chapter seven hundred and fifty-two of the laws of eighteen hundred and ninety-four, and amended by chapter five hundred and nineteen of the laws of eighteen hundred and ninety-five, chapter five hundred and sixty-two of the laws of nineteen hundred and four, chapter four hundred and seventy-two of the laws of nineteen hundred and six, chapter six hundred and seven of the laws of nineteen hundred and six, chapter five hundred and thirty-four of the laws of nineteen hundred and seven, chapter four hundred and seventy-two of the laws of nineteen hundred and eight, chapter four hundred and ninety-eight of the laws of nineteen hundred and nine, chapter eight hundred and eighty-eight of the laws of nineteen hundred and eleven and chapter five hundred and forty of the laws of nineteen hundred and thirteen,[6] is hereby amended to read as follows:

Contracts for construction, when to become operative

9. If the cost and expenses of construction of any such railroad and for acquisition of property necessary for construction and operation thereof as aforesaid shall be only partially assessed as aforesaid upon the property benefited, no provisions in any con-

---

[6] Subd. 9 was not affected by L. 1913, ch. 540.

tract for the construction thereof shall become operative until the board of estimate and apportionment or other analogous local authority shall have consented thereto and shall have prescribed a limit to the amount of city bonds, if any, available for the purpose of said contract as hereinbefore provided, and no provisions in any contract for the construction of any railroad which construction is to be paid for wholly or partly by means of local assessments shall become operative until the board of estimate and apportionment or other analogous local authority shall have levied an assessment to provide for the construction thereof, and until either assessments shall have been paid in, or assessment bonds, or in lieu thereof rapid transit construction bonds under the provisions of this section, issued by the comptroller in advance of the collection of such assessment, shall have been sold in sufficient amounts when paid for, to cover the cost and expense payable from assessments levied as aforesaid and until the board of estimate and apportionment or other analogous local authority shall have consented to such contract. In so far as any such railroad shall be constructed by means of local assessments as aforesaid, the contract for construction shall provide that any sums of money payable thereunder for or on account of such construction shall be payable only from the rapid transit construction fund of such road, and in so far as any such road shall be constructed by means of moneys appropriated by the city the contract for construction shall provide that any sums of money payable thereunder for or on account of such construction shall be payable only from the proceeds of said appropriation. In either event, the contract for construction shall provide that the city shall not be liable to any contractor for any sum or sums payable thereunder, except to the extent of moneys paid or to be paid into such rapid transit construction fund or derived or to be derived from said appropriation.

§ 6. Subdivision ten of section thirty-seven of chapter four of the laws of eighteen hundred and ninety-one, entitled "An act to provide for rapid transit railways in cities of over one million inhabitants," as added by chapter seven hundred and fifty-two of the laws of eighteen hundred and ninety-four, and amended by chapter five hundred and nineteen of the laws of eighteen hundred and ninety-five, chapter five hundred and sixty-two of the laws of nineteen hundred and four, chapter four hundred and seventy-two of the laws of nineteen hundred and six, chapter six hundred and seven of the laws of nineteen hundred and six, chapter five hundred

ch. 472,
L. 1909,
ch. 498,
and
L. 1911,
ch. 888,
amended.

and thirty-four of the laws of nineteen hundred and seven, chapter four hundred and seventy-two of the laws of nineteen hundred and eight, chapter four hundred and ninety-eight of the laws of nineteen hundred and nine, chapter eight hundred and eighty-eight of the laws of nineteen hundred and eleven and chapter five hundred and forty of the laws of nineteen hundred and thirteen,[7] is hereby amended to read as follows:

Deficiencies in fund, when paid by city.

10. In a case where the moneys collected pursuant to an assessment levied as hereinbefore provided shall be insufficient to discharge the rapid transit construction bonds so issued as aforesaid, or if the amount arising on the sale of such bonds is insufficient to pay the obligations incurred for the construction of such railroad and for acquisition of property necessary for construction and operation thereof as aforesaid, or incurred for acquisition of property, if the assessment is for cost and expense thereof separately, the deficiency up to an amount not in excess of ten per centum of the total amount of the assessment shall be paid by such city into the rapid transit construction fund, either from current revenue or from the proceeds of the sale of revenue bonds, corporate stock or other obligations of such city to be authorized and sold as provided in this act, as the board of estimate and apportionment or other such analogous local authority shall determine.

§ 7. This act shall take effect immediately.

Chap. 546.

AN ACT to amend the code of civil procedure, in relation to the appointment of court officers and attendants in the surrogate's courts of Bronx, Queens and Richmond counties.

Became a law May 8, 1915, with the approval of the Governor. Passed, three-fifths being present.

Accepted by the City.

The People of the State of New York, represented in Senate and Assembly, do enact as follows:

§ 2493 amended.

Section 1. Section twenty-four hundred and ninety-three of the code of civil procedure is hereby amended to read as follows:

§ 2493. **Appointment of court officers and attendants.** The surrogate of Kings county may appoint, and at pleasure remove,

---

[7] Subd. 10 was not affected by L. 1913, ch. 540.

"Old School" Funding Method: Transit Facilities Built by Real Estate Speculators

EXAMPLE: For Every Billion Dollars Spent On Transit, Ten Billion In New Real Estate Development Should Be Produced

City Owns Minority Interest In The Development Entity, and/or City Has Regulatory Control Over Development Entity Via "Franchise" Or "Concession"

DRAFT Guidelines Regulating Development

1. Mega Developers should pay for "flood proofing" Red Hook as per the "AECOM" plan

2. Mega Developers should pay for ongoing community educational improvements, such as "technically oriented job training", "mentoring", "after school" programs, etc.

3. Mega Developers should guarantee at least 25% affordable housing units in all new "Mega" housing development projects. Rental Rates Based Upon the Same "Sliding Rent Calculation Scale" used by NYCHA.

4. Low • Moderate and fixed income residents, and "mom and pop" businesses, should be exempted from the "land value capture tax increment'. The "tax increment' should, with certain exceptions, be entirely directed at new "Mega Development".

5. Any long time resident displaced by the project, will be offered (within a reasonable length of time) a new comparable residence in the new development, at the same rent he/she was paying previously, under the same terms, and under the same rent regulations, if any, which were in force at the time of demolition.

The Unique Genius of Hong Kong's Public Transportation System

By Neil Padukone

Passengers walk out of MTR railway carriage featuring Disney characters in the Sunny Bay station in Hong Kong. (Paul Yeung/Reuters)

New Yorkers are famous for complaining about the city's subway: despite an ever-increasing rise in fares, service never seems to get any better. And even still, ticket-sales still only funds part of the New York City subway system; the city still relies on supplementary taxes and government grants to keep trains running, as fares only cover about 45 percent of the day-to-day operating costs. Capital costs (system expansions, upgrades, and repairs) are an entirely different question, and require more state and federal grants as well as capital market bonds. And New York's system is not unique: as in other cities, New York struggles to pay existing expenses and must go into debt to pay for upgrades, that is, without raising prices.

Is this problem intractable? Not exactly. Take Hong Kong for example: The Mass Transit Railway (MTR) Corporation, which manages the subway and bus systems on Hong Kong Island and, since 2006, in the northern part of Kowloon, is considered the gold standard for transit management worldwide. In 2012, the MTR produced revenue of 36 billion Hong Kong Dollars (about U.S $5 billion) —turning a profit of $2 billion in the process. Most impressively, the farebox recovery ratio (the percentage of operational costs covered by fares) for the system was 185 percent, the world's highest. Worldwide, these numbers are practically unheard of—the next highest urban ratio, Singapore, is a mere 125 percent.

In addition to Hong Kong, the MTR Corporation runs individual subway lines in Beijing, Hangzhou, and Shenzhen in China, two lines in the London Underground, and the entire Melbourne and Stockholm systems. And in Hong Kong, the trains provide services unseen in many other systems around the world: stations have public computers, wheelchair and stroller accessibility (and the space within the train to store them), glass doors blocking the tracks, interoperable touch-and-go fare payment (which also works as a debit card in local retail), clear and sensible signage, and, on longer-distance subways, first-class cars for people who are willing to pay extra for a little leg space.

How can Hong Kong afford all of this? The answer is deceptively simple: "Value Capture."

Like no other system in the world, the MTR understands the monetary value of urban density—in other words, what economists call "agglomeration." Hong Kong is one of the world's densest cities, and businesses depend on the metro to ferry customers from one side of the territory to another. As a result, the MTR strikes a bargain with shop owners: In exchange for transporting customers, the transit agency receives a cut of the mall's profit, signs a co-ownership agreement, or accepts a percentage of property development fees. In many cases, the MTR owns the entire mall itself. The Hong Kong metro essentially functions as part of a vertically integrated business that, through a "rail plus property" model, controls both the means of transit and the places passengers visit upon departure. Two of the tallest skyscrapers in Hong Kong are MTR properties, as are many of the offices, malls, and residences next to every transit station (some of which even have direct underground connections to the train). Not to mention, all of the retail within subway stations, which themselves double as large shopping complexes, is leased from MTR.

The profits from these real estate ventures, as well as that 85 percent farebox surplus, subsidize transit development: proceeds pay for capital expansion as well as upgrades. The MTR's financial largesse means that the transit system requires less maintenance and service interruptions, which in turn reduces operating costs, streamlines capital investments, and encourages more people to use transit to get around. And more customers means more money, even if fares are relatively cheap: most commutes fall between HK $4 and HK$20 (about 50 cents to $3), depending on distance. (In London, by comparison, a Tube journey can cost as much as $18). Fare increases in Hong Kong are limited by regulations linking fares to inflation and profits, and the territory's government recently started giving a HK $600-per-month travel stipend to low-income households, defined as those earning less than HK $10,000 a month.

This model of transit management works partly because Hong Kong is a closed system: There are no suburbs from which people can commute by car, so there are strong incentives for everyone within the territory to use the system. This feature, combined with other regulations, has kept car ownership low: 6 of every 100 vehicles in Hong Kong are for personal use, whereas the number in the U.S. is closer to 70. And while the NYC subway was built over a century ago and was neglected during much of the 20th century's suburban sprawl, Hong Kong's metro was only developed in the late 1970s. As a result, it doesn't have to rely on signals technologies from the 1930s that are only slowly being upgraded (hence the track closures in New York).

As an independent corporation with the government serving as majority shareholder (rather than a public agency, ministry, or authority), the MTR has the freedom to develop real estate, to hire and fire who it will, and to take business-minded decisions—whereas other transit systems, including the one in New York, must deal with union contracts and legal restrictions. In Hong Kong, these value charges are often displaced onto consumers, causing real estate prices to go up a little faster than they otherwise might.

Still, value capture is a powerful idea for transit management. New York has tested the waters of this approach with its $2 billion 7-train extension to the Hudson Yards project, working with the state's Metropolitan Transportation Authority and the project's developers to fund the extension with property taxes from the newly served area. Dedicated taxes, too, serve a similar purpose. But fundamentally, Hong Kong's metro succeeds because it understands that a subway system is more than just a means of transportation—it is also essential to the well-being of a city's population and economy.

This article available online at:

http://www.theatlantic.com/china/archive/2013/09/the-unique-genius-of-hong-kongs-public-transportation-system/279528/

Copyright © 2015 by The Atlantic Monthly Group. All Rights Reserved.

www.ingramcontent.com/pod-product-compliance
Lightning Source LLC
Chambersburg PA
CBHW080921170526
45158CB00008B/2191